ENCYCLOPEDIA OF
MAMMALS

ENCYCLOPEDIA OF
MAMMALS

CONSULTANT EDITORS
Dr Edwin Gould

Curator Emeritus, National Zoological Park,

Smithsonian Institution, Washington DC, USA

&

Dr George McKay

Honorary Associate, School of Biological Sciences,

Macquarie University, Sydney, Australia

ILLUSTRATIONS BY
Dr David Kirshner

Published in the United States by
Academic Press, A Division of Harcourt Brace & Company
525 B Street, Suite 1900, San Diego, CA 92101-4495, USA
Distributed worldwide exclusively by Academic Press
(except in Australia and New Zealand)

Senior Editor, Life Sciences: Charles R. Crumly Ph.D.

Conceived and produced by Weldon Owen Pty Limited
59 Victoria Street, McMahons Point, NSW 2060, Australia
A member of the Weldon Owen Group of Companies
Sydney • San Francisco

First published in 1990
Second edition 1998
Copyright © Weldon Owen Pty Limited 1998

Publisher: Sheena Coupe
Associate Publisher: Lynn Humphries
Project Editors: Jenni Bruce; Alison Pressley
Editorial Assistant: Veronica Hilton
Captions: Tom Grant; Colin Groves; David Kirshner; Ronald Strahan
Index: Garry Cousins
Designers: Sue Rawkins; Anita Rowney
Picture Research: Annette Crueger
Maps and Diagrams: Alistair Barnard; Sue Rawkins
Illustrations of Extinct Species: Alistair Barnard
Production Manager: Caroline Webber
Production Assistant: Kylie Lawson
Vice President International Sales: Stuart Laurence

Co-ordination of scientific and editorial contributors by
Linda Gibson, Project Manager, Australian Museum Business Services

ISBN 0-12-293670-1

A catalog record for this book is available from
the Library of Congress, Washington, DC.

Printed by Kyodo Printing Co. (Singapore) Pte Ltd
Printed in Singapore

A WELDON OWEN PRODUCTION

Endpapers: The puma, also known as the cougar, mountain lion, or panther, is an endangered species. Photo by Leonard Lee Rue III/Bruce Coleman Ltd
Page 1: An ermine or stoat in its summer coat of chestnut with a white underside. In winter, the entire coat turns white. Photo by Hans Reinhard/Bruce Coleman Ltd
Pages 2–3: Tail of a humpback whale seen through ice. Photo by Tim Davis/Tony Stone Images
Pages 4–5: The Barbary macaque is better known as the "Barbary ape".
Page 7: The gaping mouth of a yawning hippopotamus. Photo by Bob Campbell/Bruce Coleman Ltd
Pages 10–11: Wild boar forage for food in family parties. Photo by Hans Reinhard/Bruce Coleman Ltd
Pages 12–13: Herds of impala feed on the rich short grass of the wet season. Photo by Jonathan Scott/Planet Earth Pictures
Pages 46–47: Meerkats warm themselves in the morning sun. Photo by David Macdonald/Oxford Scientific Films

Jean-Paul Ferrero / Auscape International

CONSULTANT EDITORS

Dr Edwin Gould
Formerly Curator of Mammals, National Zoological Park,
Smithsonian Institution, Washington DC, USA

Dr George McKay
Honorary Associate, School of Biological Sciences,
Macquarie University, Sydney, Australia

CONTRIBUTORS

Dr M. M. Bryden
Professor of Veterinary Anatomy,
University of Sydney,
Australia

Dr M. J. Delany
Emeritus Professor of Environmental Science,
University of Bradford,
England

Dr Valerius Geist
Professor Emeritus,
University of Calgary,
Canada

Dr Tom Grant
Honorary Visiting Fellow,
School of Biological Science,
University of New South Wales,
Australia

Dr Cclin Groves
Reader in Anthropology,
Australian National University

J. E. Hill †
formerly Principal Scientific Officer,
Mammal Section,
British Museum (Natural History),
London

Dr Tom Kemp
Curator of the Zoological Collections,
University Museum,
Oxford

Judith E. King
formerly of the Osteology Section,
British Museum (Natural History),
London

Dr Gordon L. Kirkland, Jr
Director of the Vertebrate Museum and
Professor of Biology,
Shippensburg University,
Pennsylvania, USA

Dr Anne LaBastille
Consultant Lecturer,
Photographer, and Author

Dr Helene Marsh
Professor in Zoology,
James Cook University,
Townsville, Australia

Dr Norman Owen-Smith
Reader,
Department of Zoology,
University of Witwatersrand,
South Africa

Dr Jeheskel (Hezy) Shoshani
Research Associate,
Cranbrook Institute of Science, USA, and
President,
Elephant Research Foundation, USA

Dr D. Michael Stoddart
Deputy Vice Chancellor,
University of New England,
Australia

Dr R. David Stone
Environmental Consultant,
Conservation Advisory Services,
Switzerland

Ronald Strahan
Research Associate,
The Australian Museum,
Sydney

Dr W. Chris Wozencraft
Assistant Professor in Biology,
Lewis-Clark State College,
Oregon, USA

CONTENTS

INTRODUCTION   10

PART ONE
THE WORLD OF MAMMALS

INTRODUCING MAMMALS   14
George McKay

CLASSIFYING MAMMALS   19
Original text by J. E. Hill;
revised by George McKay

MAMMALS THROUGH THE AGES   22
Tom Kemp

HABITATS & ADAPTATION   28
Gordon L. Kirkland, Jr

MAMMAL BEHAVIOR   33
D. Michael Stoddart

ENDANGERED SPECIES   38
Original text by Anne LaBastille;
revised by George McKay

PART TWO
KINDS OF MAMMALS

MONOTREMES   48
Tom Grant

MARSUPIALS   52
Ronald Strahan

ANTEATERS, SLOTHS   66
& ARMADILLOS
R. David Stone

INSECTIVORES   74
R. David Stone

TREE SHREWS   86
R. David Stone

FLYING LEMURS   88
Original text by J. E. Hill;
revised by George McKay

BATS   90
Original text by J. E. Hill;
revised by George McKay

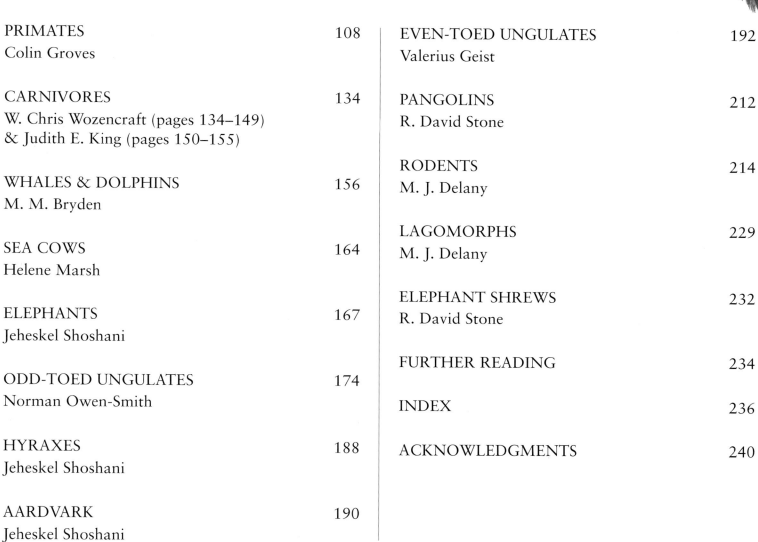

PRIMATES   108
Colin Groves

CARNIVORES   134
W. Chris Wozencraft (pages 134–149)
& Judith E. King (pages 150–155)

WHALES & DOLPHINS   156
M. M. Bryden

SEA COWS   164
Helene Marsh

ELEPHANTS   167
Jeheskel Shoshani

ODD-TOED UNGULATES   174
Norman Owen-Smith

HYRAXES   188
Jeheskel Shoshani

AARDVARK   190
Jeheskel Shoshani

EVEN-TOED UNGULATES   192
Valerius Geist

PANGOLINS   212
R. David Stone

RODENTS   214
M. J. Delany

LAGOMORPHS   229
M. J. Delany

ELEPHANT SHREWS   232
R. David Stone

FURTHER READING   234

INDEX   236

ACKNOWLEDGMENTS   240

INTRODUCTION

Today, mammals come in all sizes and shapes, from tiny shrews and bats that weigh 2 grams to enormous blue whales that tip the scales at 140 tonnes. And they occupy niches everywhere—land, sea, and air. But their success was a long time in coming. Mammals first appeared as mouse-sized insect eaters, descended from therapsid reptiles during the Triassic period 245 million years ago, at a time when the major land masses were still connected. During the Cretaceous period, dinosaurs suddenly disappeared from the Earth and mammals took their place. Ecological niches for plant- and meat-eaters as well as omnivores of all sizes became available. Then, when evergreens such as conifers and cycads began to be replaced by flowering plants with edible fruits and seeds, such as palms, magnolias, and tulip poplars, mammals evolved. Seed dispersal and pollination eventually became a part of that interplay.

Mammals developed new, complex behaviors that forever changed communication and parental and social behavior, and ultimately enabled the passing of information from one generation to the next. The new mammary glands, able to produce and secrete milk, provided the opportunity for maternal care and as a result bonding, the seed of social behavior, occurred between mother and offspring.

To human beings, mammals have taken on much greater significance than their biological places in the ecosystem might indicate. The great animals we associate with the vitality of life on Earth—the elephants, apes, bears, lions, tigers—thread through our literature and act as symbols of power, stability, longevity, wisdom, memory. This major class of vertebrates also supports human life on Earth as pets, domestic work animals, and food.

Common as mammals are, many species are vulnerable to extinction—an outcome that has been as much a part of evolution as biological change. But with the rise of the most successful predators of all time—human beings—species are disappearing at a rate that far exceeds the evolution of new species. Although many species of mammals disappeared at the hands of hunters during the Pleistocene, many more are becoming extinct today. Habitat destruction in the midst of the current human population explosion makes life for many species impossible. Sadly, some of the species mentioned in this encyclopedia may be extinct within a decade. Nevertheless, these pages should give every reader an appreciation of the structural and ecological diversity as well as the behavioral richness of mammals. Perhaps as we begin to respect and appreciate mammals more, we will be able to devise ways to conserve their natural habitats. Then we can continue to live together and share as fellow beings our beautiful planet through the next era.

EDWIN GOULD
Consultant Editor

PART ONE
THE WORLD

OF MAMMALS

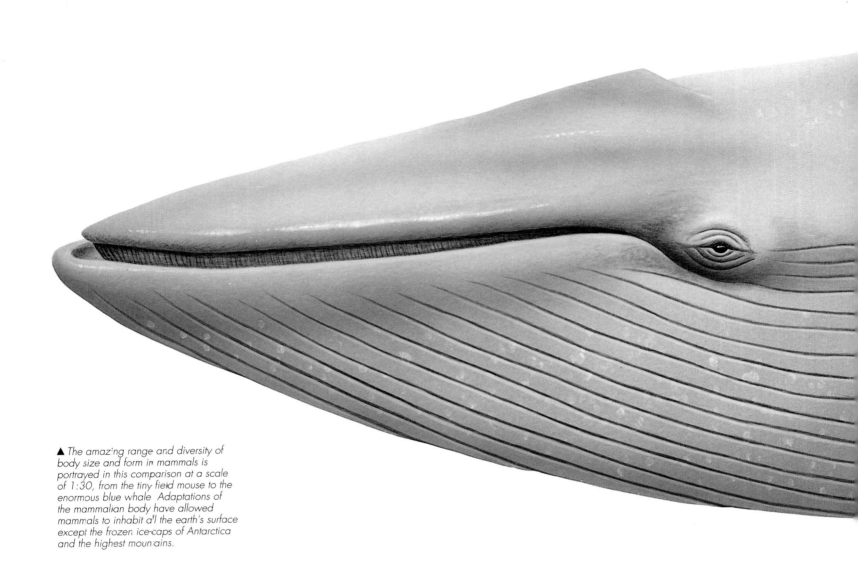

▲ The amazing range and diversity of body size and form in mammals is portrayed in this comparison at a scale of 1:30, from the tiny field mouse to the enormous blue whale Adaptations of the mammalian body have allowed mammals to inhabit all the earth's surface except the frozen ice-caps of Antarctica and the highest mountains.

INTRODUCING MAMMALS

S cientists divide the animal kingdom into several major groups for classification purposes. By far the largest group is the invertebrates: it contains about 95 percent of the millions of known species of animals, including sponges, mollusks, arthropods, and insects. Groups of vertebrates, or animals with backbones, contain the other 5 percent of known species. They can be divided roughly into fishes, amphibians, reptiles, birds, and mammals. The class Mammalia consists of fewer than 5,000 species, but the sheer diversity of mammals is astonishing. From tiny field-mice to the mighty blue whale, from a hippopotamus to a bat, from armadillos to gorillas, the class Mammalia encompasses some of the best-known and most-studied, as well as some of the least-known, members of the animal kingdom. It also includes human beings. We are mammals, classed along with monkeys, lemurs, and apes in the order Primates.

FEATURES THAT LINK MAMMALS

Living mammals are warm-blooded animals that suckle their young on milk and have a body covering of hair or fur, prominent external ears, and a mouth armed with teeth. These features—which, incidentally, are not even shared by all mammals—serve to differentiate mammals from all other living vertebrates. For scientists, however, the single feature that defines mammals is the method by which the dentary bone of the lower jaw—the one that houses the teeth—articulates directly with the skull.

Mammals evolved from a group of carnivorous reptiles, and the most primitive mammals were, like their ancestors, small flesh-eaters. As they evolved, mammals spread and adapted to many different habitats and now occupy a wide variety of niches, from small flying insect-eaters to large terrestrial grazers, and from tiny burrowing flesh-eaters to the largest of living animals, the plankton-feeding whales.

SKULLS, JAWBONES, AND TEETH

The mammal-like reptiles from which mammals evolved are known as the therapsids. Many changes in the skull occurred during the evolution of the therapsids, but the most important concerned the jaws. Early during therapsid evolution the upper jaw became firmly attached to the rest of the skull and lost the independent mobility that can still be seen in such reptiles as

A TYPICAL MAMMAL'S HAIR

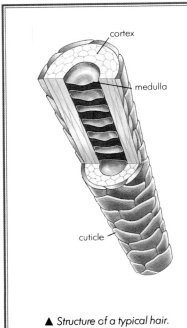

Each hair consists of three concentric layers of cells, which develop in the follicle and then die as they are pushed outward. The innermost layer, called the medulla, is made up of the dead remnants of cells that contain air. This increases the insulating ability of the fur and in some cases provides buoyancy when the animal is swimming. Very thin hairs often lack a medulla, as do the wool fibers of sheep.

Surrounding the medulla are the cells of the cortex, which become invaded with fibrils of keratin and which may also contain pigment granules. The outer layer is the cuticle, the individual cells of which resemble microscopic scales. The cuticle cells and medulla cells show a wide variety of patterns which frequently differ between hair types and between species, so a species can often be identified from the structure of its hair.

▲ *Structure of a typical hair.*

As teeth are the major food-gathering structures of most mammals, they have been subject to a great deal of evolutionary change and have developed a number of distinctive features. They have, for example, different shapes in different parts of the jaw; they are replaced only once, if at all, and not many times as in fish and reptiles; and they are anchored in sockets in the bone.

SKIN AND FUR

The skin of mammals shows a number of characteristics not found in other backboned animals. These include the continuous growth and replacement of the epidermis, the outermost of the three layers of skin; the presence of hairs; and a number of new types of skin glands. The waterproof epidermis houses the hairs and skin glands, but contains no other structures. The second layer, or dermis, contains supporting collagen fibres, blood vessels, a variety of sensory nerve endings, and muscles that can cause the hairs to become erect. Attached to the inner boundary of the dermis is the sub-dermal fatty layer, which provides insulation.

Hairs, which vary greatly in shape and size, grow from the base of follicles and have a complex structure. Many mammals have at least two main hair types in their body covering: long guard hairs, and shorter hairs or underfur. The combination of these two hair types provides efficient thermal insulation by trapping a layer of still air close to the skin. In some large mammals, hair is sparse or absent, and these species rely on the insulating ability of the skin. Each individual hair is subject to wear and has a limited usable life. Replacement of hairs, or molting, may occur continuously or, in colder climates, at specific times of the year. Some species, such as the ermine, have separate summer and winter coats.

Specialized hairs, or "vibrissae", occur at various points on the body, usually on the head but also on the forearms and feet. Vibrissae have specific sensory cells associated with them and provide important tactile information.

Of the different types of glands occurring in the skin, the most important are the mammary glands. These secrete the milk that nourishes the young. Mammary glands may occur in both sexes in placental mammals but are functional only in the female. Females of all mammals possess mammary glands, but projecting teats or nipples occur only in the marsupials and placentals; in the monotremes the glands open in an areolar area which the young can nuzzle with their short beaks.

Sebaceous glands are present in most mammals, in close association with the hair follicles. They produce an oily secretion that serves to lubricate and protect the hairs. Many mammals have sweat glands, which provide evaporative cooling and elimination of metabolic wastes. Sweat glands may be widely distributed over the body, as in humans,

snakes. Later, there were significant changes to the composition of the lower jaw: the bones housing the teeth became greatly enlarged and the other bones became smaller, until the dentary became the only bone in the lower jaw. In the earliest mammals, the dentary bones achieved direct contact with the squamosal bone of the skull.

One of the important developments in the mammal-like reptiles was the reduction in the number of bones in the skull and the strengthening of those that remained. The bones around the eye became less important while those surrounding the brain enlarged to accommodate an increasingly large nervous system.

▼ *Like most mammals living in cold climates, the mink has an insulating coat of soft, close-packed hair. For thousands of years, we have used the fur of such animals as insulation for our own almost naked bodies.*

or restricted to particular areas. They may even be completely lacking. Scent glands are highly specialized structures that produce volatile odorous secretions. These are important as a means of communicating information; in some species, they are important as a means of self-defense.

BACKBONES AND LIMBS

Living mammals have a double contact between the skull and the atlas—the first cervical (neck) vertebra. Most living mammals have seven cervical vertebrae, although this number does vary among sloths and is reduced to six in manatees. In most mammals, ribs are associated only with the chest vertebrae; the exceptions are the monotremes, which have cervical ribs.

The limbs of mammals have retained the basic vertebrate five-fingered plan with few modifications, even in such diverse groups as bats, whose hands have become wings, and whales, whose forelimbs are oar-like. In the hoofed mammals, there has been a progressive reduction in the number of toes from five to two, or even one. The growth of limb bones in mammals differs from that of reptiles in that growth is not indefinite and is confined to two growing points near either end of each long bone.

Monotremes, like a number of early mammal groups, have retained a primitive shoulder girdle with a number of separate bones, but the other living mammals have a shoulder girdle which consists of a scapula (shoulder-blade) and a clavicle (collar-bone). The clavicle is frequently lost in such groups as hoofed mammals, which have evolved extreme flexibility of the limbs to increase their running speed, but is retained and indeed strengthened in those primates that rely on their strong arms for climbing.

The pelvic girdles of mammals consist basically of three fused bones, as in most reptiles, although those of the monotremes and marsupials have an additional bone which provides for the insertion of muscles supporting the ventral body wall.

BRAIN AND SENSES

Mammals have much larger brains, relative to body size, than other vertebrates. The increased brain size is due to an expansion of the cerebral hemispheres. A new component of the cerebrum, the neopallium, which occurs as a very small region in some reptiles, has expanded to provide a covering of gray cells over the ancestral reptilian brain. While all mammals possess a neopallium, the placentals also have a connection between the two cerebral hemispheres: the corpus callosum.

The sense of smell is highly developed in most mammals, although it is less important in many primates and is greatly reduced or absent in cetaceans. The sense of taste, however, is relatively unspecialized. Touch receptors occur widely on the body but are especially important in association with the specialized hairs called the vibrissae.

The mammalian eye is well developed and is basically similar in structure and function to the reptilian eye. Color vision has arisen independently in several mammals, including primates and some rodents, and binocular vision, which permits efficient estimation of distance, is particularly well developed in primates. Many nocturnal mammals have a reflective layer at the back of the eye which increases visual acuity by reflecting weak incoming light back onto the retina.

Hearing is important to most mammals and a number of specializations have evolved, particularly the amplifying function of the middle ear. Most mammals have an external ear that collects sound and concentrates it on the opening to the middle ear. Many mammals can hear sounds of very high pitch, and this has been exploited—by bats, for instance—for echolocation, by which the animal detects obstacles by listening to the reflections of its own sound pulses.

HOW MAMMALS HEAR

The mammalian middle ear is a complex series of membranes and bones that transmit and amplify sound from the outside to the auditory cells in the inner ear. Sound waves impinge on the tympanic membrane, supported by the tympanic bone, and are transmitted via the malleus, incus, and stapes to the oval window of the inner ear. Here they set up an enhanced series of compression waves in the fluid of the inner ear, which in turn excite the sensory cells of the auditory epithelium. An interesting evolutionary phenomenon is the transfer of function of two former articulating bones (the malleus and the incus) from controlling the lower jaw movement to assisting and amplifying hearing.

In the more primitive mammals this middle ear structure lies outside the skull, but in more specialized forms it is protected by a bony covering, or bulla, to which the tympanic bone becomes fused.

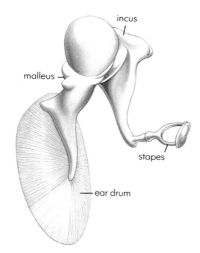

◀ The three bones in a mammal's ear: the malleus, incus, and stapes. Two of these, the malleus and incus, form part of the jaw hinge in reptiles.

incus
malleus
stapes
ear drum

David Macdonald/Oxford Scientific Films

▲ A female meerkat suckling young. Mammals are named after the mammary glands, found only in this group of animals. Milk is a complete food, providing all the nutrients the young need until they are ready to eat by themselves.

scrotum, where the sperm can develop at a temperature lower than normal body temperature. After copulation and ejaculation, the sperm migrate rapidly to the upper portion of the female fallopian tubes, where they fertilize the ova.

In monotremes the fertilized egg is surrounded in the uterus by a shell membrane and a leathery shell. It is then incubated in the uterus for about two weeks before being laid and further incubated externally. In marsupials a shell membrane, but no shell, is deposited. In placental mammals shell membranes are absent.

The young of marsupials are born in a very undeveloped state after a very short gestation. They attach firmly to the teat of a mammary gland which may or may not be located in a pouch on the mother's belly. In placental mammals gestation is relatively longer and the newborn young are in a more advanced state of development, although in some species they are still naked and helpless. A major trend in the evolution of placental mammals has been the lengthening of the gestation period and the birth of more highly developed young. This is usually associated with the production of fewer young in each litter.

OTHER ORGAN SYSTEMS

In mammals, a muscular diaphragm separates the chest and abdominal cavities. Contraction of the diaphragm, together with the action of the muscles connecting the ribs, draws air into the lungs. The upper respiratory system consists of a trachea which branches out to form the bronchi, leading to the lungs. At the upper end of the trachea is the larynx, where sound is produced.

The circulatory system is dominated by a four-chambered heart, which has a different arrangement from the four-chambered hearts of crocodiles and birds. Separation of pulmonary and systemic circulations is complete in the adult. The red blood cells bind and transport oxygen and carbon dioxide from and to the lungs. A variety of white blood cells perform the housekeeping functions of the body's immune system. All mammals can maintain their body temperature above that of their surroundings, and many are also capable of lowering their body temperature—either on a daily basis or for a prolonged period of torpor or hibernation.

Mammalian kidneys are "kidney-shaped" rather than elongated structures as in other vertebrates. Like many fish, amphibians, and turtles, mammals excrete nitrogenous wastes as urea rather than as uric acid, as most reptiles and all birds do.

The digestive system shows a number of specializations in mammals, particularly among those groups that ferment plant material in a highly modified stomach or in a cecum, which is a sac occurring at the junction of the small intestine and the proximal colon.

GEORGE MCKAY

HOW MAMMALS REPRODUCE

Mammalian reproduction is characterized by internal fertilization which results in an amniote egg. The female reproductive system consists of a pair of ovaries and their associated ducts. Under the influence of hormones produced by the pituitary gland, mature ova are released into the fallopian tubes. The control of ovulation may be environmental, as in many highly seasonal breeders, or it may be induced by copulation, as in cats. Males produce sperm in the testes, which are located within the body cavity or outside in a

CLASSIFYING MAMMALS

There are several million different kinds of plants and animals. The purpose of classification is to provide each of these with a unique name by which it may be known; to describe it so that it may be recognized; and to place it in a formal arrangement or hierarchy that will express its relationship to others. The science of arranging plants and animals in this way is called taxonomy or sometimes systematics.

VERNACULAR NAMES
Many mammals have vernacular or common names, although many do not. These are not used in classification because they may differ from country to country, or from one language to another, and can be ambiguous.

SCIENTIFIC NAMES
Scientific names are always in Latin, partly because this was the common language of the early naturalists, but chiefly because Latin is universally accepted by scientists as a means of avoiding the problems of translation and ambiguity.

The scientific name of any kind or species of plant or animal always consists of two words: the first or generic name indicates the relationship of the species to others, the second or specific name signifies the particular species concerned.

Scientific names are often descriptive, drawing attention to some significant feature of their subject, or they may refer to the country or place whence it originated. Sometimes names commemorate a particular person, and some are classical in origin, but they can also be quite arbitrary.

SPECIES
The species is the basic category in classification. About 4,600 species of modern mammals are recognized and although some, such as the giant panda, the lion, or the tiger, are easily seen to be distinct, the majority are much less obviously so and are defined by differences in external form and structure, in the structure of the skull and teeth, or even in such attributes as chromosome pattern and structure. The interpretation of similarities and differences is often subjective, and for this reason the total number of mammalian species varies slightly from one authority to another. Also, from time to time new species are discovered and described, even today.

GENERA
Species with a number of features in common are grouped together in genera, although occasionally a very strongly characterized species may be the sole member of its own genus. Thus the genus *Vulpes* includes a number of species of fox besides the red or common fox *Vulpes vulpes,* but the Arctic fox is considered sufficiently distinct from these other foxes to be placed by itself in a separate genus as *Alopex lagopus.*

Genera are based on more far-reaching characters than species and are sometimes divided into subgenera to emphasize the features of a particular species or group of species. Again, the interpretation of their characters is subjective and the number of modern genera of mammals is not irrevocably fixed: currently about 1,130 are listed.

FAMILIES
In the same way that genera signify groups of species, or sometimes one strongly characteristic species, so families comprise groups of genera that share common characteristics or similarities: a family may include only one very characteristic genus, on occasion with only a single species. There are 136 families of modern mammals.

ORDERS
Families that share major characteristics are grouped into orders. As a rule these are quite distinctive and separated from others by obvious features, but sometimes the distinction may be less immediately apparent. For example, elephants occupy an order of their own, the Proboscidea, which can be easily recognized, but the distinction between tree shrews (Scandentia) and elephant shrews (Macroscelidea) is less obvious.

Most orders include several or many families, but a number besides the Proboscidea contain but a single family. One order, the Tubulidentata, has only a single genus and species, the African aardvark or ant bear *Orycteropus afer.* Orders may be subdivided into suborders to emphasize differences among their families. There are 20 orders of modern mammals, the Rodentia or rodents being the largest, with about 40 percent of known mammal species, followed by the Chiroptera or bats, which have about a further 23 percent.

THE RED FOX: AN EXAMPLE OF MAMMAL CLASSIFICATION

The basis of mammalian classification can be illustrated by taking as an example the red or common fox *Vulpes vulpes* of northern Europe, Asia, and northern America. The genus *Vulpes* contains besides this species several other species of fox, and is grouped with other genera of foxes, and with dogs, wolves, and jackals in the family Canidae. Together with a number of other families of carnivorous mammals, such as the Ursidae (bears and pandas), Procyonidae (raccoons), Hyaenidae (hyenas), and Felidae (cats), the Canidae are part of the order Carnivora, which in turn belongs to the infraclass Eutheria, and the subclass Theria.

Class: Mammalia
Infraclass: Eutheria
Subclass: Theria
Order: Carnivora
Family: Canidae
Genus: *Vulpes*
Species: *Vulpes vulpes*

▼ *Red fox cubs of the species* Vulpes vulpes. *Classification enables us to catalogue animals conveniently and serves to indicate the degree of evolutionary relationship between different species.*

Hans Reinhard/Bruce Coleman Ltd

HIGHER CLASSIFICATION

Mammals belong to the class Mammalia, in which two major divisions are recognized. One, the Prototheria, includes only the spiny anteaters of Australia and New Guinea and the Australian duck-billed platypus, which are unique among mammals in that they reproduce by laying eggs. The other, the Theria, includes all other modern mammals, but itself has two divisions. One, the Metatheria, is reserved for the order Marsupialia, or marsupials, whose young are born in a very undeveloped state and complete their development in a pouch or fold on the mother's abdomen. The remainder of modern mammals belong to the second group, the Eutheria, whose young develop to a relatively advanced stage within the body of the mother.

FOSSIL MAMMALS

Modern mammals constitute a very small part of the total kinds that have existed since mammals first evolved 200 million years ago. The most recent fossils may represent genera or even species known today, and many fossil mammals belong to the same orders and families as their modern counterparts. Many fossils, however, represent totally different mammals that have evolved, thrived, and then become extinct during the long period since mammals first appeared.

CHANGES IN CLASSIFICATION

New ways of assessing the similarities and differences between mammals, or of extending the range of characters upon which classification depends beyond those used traditionally, lead from time to time to changes in classification. Also, the interpretation and meaning of their features may differ from one authority to another. There is no immutable classification, and in some instances there is not complete agreement as to the relationships among and between species, genera, and even families and orders.

Several authors have proposed dividing the marsupials into as many as seven separate orders and these have been accepted by some texts including Wilson & Reeder, 1993. They have not yet been universally accepted, however. As the different lines of evidence point to different interpretations of relationships, yet all marsupials are their own closest relatives, we prefer to follow the conservative course of listing a single order.

Rodents may possibly represent two independent lineages. Even more upsetting would be to place the whales in the order Artiodactyla between pigs and hippos as DNA studies suggest. The unique adaptations of whales to a pelagic life warrant the retention of the order Cetacea. The classifications in the two major references differ in many interpretations. The classification we present here is a compromise erring on the side of caution.

J. E. HILL & GEORGE MCKAY

ORDERS AND FAMILIES OF MAMMALS

The following list is based on *A World List of Mammalian Species* by G. B. Corbet and J. E. Hill, 3rd edition, 1991, Facts on File Publications, New York/British Museum (Natural History), London. Some adjustments have been made within orders in accordance with each author's preferences.

CLASS MAMMALIA

Subclass Prototheria

ORDER
MONOTREMATA — MONOTREMES
Tachyglossidae — Spiny anteaters
Ornithorhynchidae — Duck-billed platypus

Subclass Theria

Infraclass Metatheria

ORDER MARSUPIALIA — MARSUPIALS
Didelphidae — American opossums
Microbiotheriidae — Colocolos
Caenolestidae — Shrew-opossums
Dasyuridae — Marsupial mice, etc.
Myrmecobiidae — Numbat
Thylacinidae — Thylacine
Notoryctidae — Marsupial mole
Peramelidae — Bandicoots
Peroryctidae — Spiny bandicoots
Vombatidae — Wombats
Phascolarctidae — Koala
Phalangeridae — Phalangers
Petauridae — Gliding phalangers
Pseudocheiridae — Ringtail possums
Burramyidae — Pygmy possums
Acrobatidae — Feathertails
Tarsipedidae — Honey possum
Macropodidae — Kangaroos, wallabies
Potoroidae — Bettongs

Infraclass Eutheria

ORDER XENARTHRA — ANTEATERS, SLOTHS & ARMADILLOS
Myrmecophagidae — American anteaters
Bradypodidae — Three-toed sloths
Megalonychidae — Two-toed sloths
Dasypodidae — Armadillos

ORDER INSECTIVORA — INSECTIVORES
Solenodontidae — Solenodons
Tenrecidae — Tenrecs, otter shrews
Chrysochloridae — Golden moles
Erinaceidae — Hedgehogs, moonrats
Soricidae — Shrews
Talpidae — Moles, desmans

ORDER SCANDENTIA — TREE SHREWS
Tupaiidae — Tree shrews

ORDER DERMOPTERA — FLYING LEMURS, COLUGOS
Cynocephalidae — Flying lemurs, colugos

ORDER CHIROPTERA — BATS
Pteropodidae — Old World fruit bats
Rhinopomatidae — Mouse-tailed bats
Emballonuridae — Sheath-tailed bats
Craseonycteridae — Hog-nosed bat, bumblebee bat
Nycteridae — Slit-faced bats
Megadermatidae — False vampire bats
Rhinolophidae — Horseshoe bats
Hipposideridae — Old World leaf-nosed bats
Noctilionidae — Bulldog bats
Mormoopidae — Naked-backed bats
Phyllostomidae — New World leaf-nosed bats
Natalidae — Funnel-eared bats
Furipteridae — Smoky bats
Thyropteridae — Disc-winged bats
Myzopodidae — Old World sucker-footed bat
Vespertilionidae — Vespertilionid bats
Mystacinidae — New Zealand short-tailed bats
Molossidae — Free-tailed bats

ORDER PRIMATES — PRIMATES
Cheirogaleidae — Dwarf lemurs
Lemuridae — Large lemurs
Megaladapidae — Sportive lemurs
Indridae — Leaping lemurs
Daubentoniidae — Aye-aye
Loridae — Lorises, galagos
Tarsiidae — Tarsiers
Callitrichidae — Marmosets, tamarins
Cebidae — New World monkeys
Cercopithecidae — Old World monkeys
Hylobatidae — Gibbons
Hominidae — Apes, man

ORDER CARNIVORA — CARNIVORES
Canidae — Dogs, foxes
Ursidae — Bears, pandas
Procyonidae — Raccoons, etc.
Mustelidae — Weasels, etc.
Viverridae — Civets, etc.
Herpestidae — Mongooses
Hyaenidae — Hyenas
Felidae — Cats
Otariidae — Sealions
Odobenidae — Walrus
Phocidae — Seals

ORDER CETACEA — WHALES, DOLPHINS
Platanistidae — River dolphins
Delphinidae — Dolphins
Phocoenidae — Porpoises
Monodontidae — Narwhal, white whale
Physeteridae — Sperm whales
Ziphiidae — Beaked whales
Eschrichtiidae — Gray whale
Balaenopteridae — Rorquals
Balaenidae — Right whales

ORDER SIRENIA — SEA COWS
Dugonidae — Dugong
Trichechidae — Manatees

ORDER PROBOSCIDEA — ELEPHANTS
Elephantidae — Elephants

ORDER PERISSODACTYLA — ODD-TOED UNGULATES
Equidae — Horses
Tapiridae — Tapirs
Rhinocerotidae — Rhinoceroses

ORDER HYRACOIDEA — HYRAXES
Procaviidae — Hyraxes

ORDER TUBULIDENTATA — AARDVARK
Orycteropodidae — Aardvark

ORDER ARTIODACTYLA — EVEN-TOED UNGULATES
Suidae — Pigs
Tayassuidae — Peccaries
Hippopotamidae — Hippopotamuses
Camelidae — Camels, llamas
Tragulidae — Mouse deer
Moschidae — Musk deer
Cervidae — Deer
Giraffidae — Giraffe, okapi
Antilocapridae — Pronghorn
Bovidae — Cattle, antelopes, etc.

ORDER PHOLIDOTA — PANGOLINS, SCALY ANTEATERS
Manidae — Pangolins, scaly anteater

ORDER RODENTIA — RODENTS
Aplodontidae — Mountain beaver
Sciuridae — Squirrels, marmots, etc.
Geomyidae — Pocket gophers
Heteromyidae — Pocket mice

Castoridae — Beavers
Anomaluridae — Scaly-tailed squirrels
Pedetidae — Spring hare
Muridae — Rats, mice, gerbils, etc.
Gliridae — Dormice
Seleviniidae — Desert dormouse
Zapodidae — Jumping mice
Dipodidae — Jerboas
Hystricidae — Old World porcupines
Erethizontidae — New World porcupines
Caviidae — Guinea pigs, etc.
Hydrochaeridae — Capybara
Dinomyidae — Pacarana
Dasyproctidae — Agoutis, pacas
Chinchillidae — Chinchillas, etc.
Capromyidae — Hutias, etc.
Myocastoridae — Coypu
Octodontidae — Degus, etc.
Ctenomyidae — Tuco-tucos
Abrocomidae — Chinchilla-rats
Echymidae — Spiny rats
Thryonomyidae — Cane rats
Petromyidae — African rock-rat
Bathyergidae — African mole-rats
Ctenodactylidae — Gundis

ORDER LAGOMORPHA — LAGOMORPHS
Ochotonidae — Pikas
Leporidae — Rabbits, hares

ORDER MACROSCELIDEA — ELEPHANT SHREWS
Macroscelididae — Elephant shrews

▲ One of the earliest and most primitive of the "mammal-like reptiles", Dimetrodon grew to about 3 meters in length. It inhabited parts of what is now North America 300 million years ago.

MAMMALS THROUGH THE AGES

The mammals of the world today are the survivors of a long history that started about 195 million years ago. In rocks of that age, the very first unmistakable mammal fossils occur, as tiny insect-eating animals that look a little like shrews. For almost the first two-thirds of their subsequent history, mammals remained small, inconspicuous animals, probably active only at night. During all this time — the Jurassic and Cretaceous periods of the geologists' time chart — they shared their habitat with the dinosaurs. But when the great extinction of the dinosaurs occurred at the end of the Cretaceous, the land appears to have been opened up for exploitation by mammals. From that moment on, throughout the following 65 million years to the present, many different kinds of mammals, large and small, carnivorous and herbivorous, terrestrial and aquatic, have evolved, flourished, and disappeared, to be replaced by yet newer kinds.

THE ORIGIN OF MAMMALS

Mammals fossilize relatively easily, because their skeletons are generally robust and withstand well the rigors of the fossilization processes. The teeth and jaws in particular are often perfectly preserved, which is doubly fortunate because the exact structure of any particular mammal's teeth tells a great deal about the evolution and biology of that mammal. Consequently the history of the mammals is better known than that of any other comparably sized group of organisms.

Three hundred million years ago the land was populated by primitive amphibians and reptiles, living in and around the extensive wet tropical swamps of the time. Among the reptile fossils, remains of a few forms that were rather larger than the rest, and which had a pair of windows, or temporal fenestrae, in the hind part of their bony skulls, have been found. These windows are still found in a modified way in mammals, indicating that these animals were the first of the "mammal-like reptiles", or synapsids as they are correctly called. The importance of this group is that it includes the animals from which mammals eventually evolved. The early members of the group were not, however, very like mammals, for they still had simple teeth with weak jaws, and clumsy, sprawling limbs; certainly they were not warm-blooded. A well-known example of the early, primitive synapsids is *Dimetrodon,* the finback reptile, which lived over a wide area of what is now North America.

In due course, more advanced kinds of mammal-like reptiles called the Therapsida evolved from *Dimetrodon*-like ancestors. Therapsids had developed more powerful jaw-closing muscles, which they could use with a variety of more elaborate dentitions such as very enlarged canine teeth, or even horny beaks. They also possessed longer, more slender limbs which must have allowed them to run more rapidly and with greater agility in pursuit of their prey, or to escape from predators. Therapsids had probably become more warm-blooded and larger brained as well, although these things are more difficult to tell from fossils.

One particular group of advanced therapsids, called the Cynodontia, had evolved several other distinctly mammalian characteristics. They had much more mammalian-like teeth, with several cusps, which worked by a shearing and crushing action between the upper and the lower teeth. Along with the modifications to the teeth themselves, very powerful yet also very accurate jaw muscles had to evolve to give these teeth greater biting power. This involved a great enlargement of the bone that carried the lower teeth, the dentary bone, so that it could also carry the attachments of the lower jaw muscles. The other bones of the lower jaw, called the postdentary bones, were correspondingly reduced and weakened. In the cynodonts, these postdentary bones had adopted a new function. Along with the old hinge bone or quadrate of the skull, they transmitted sound waves from a rudimentary ear drum, attached to the jaw, to the auditory region of the animal's braincase. In this way, the cynodonts show an approach towards the fully mammalian arrangement where only the dentary bone is left in the lower jaw, and it has

▼ *Cynognathus was a carnivorous cynodont, an advanced version of the "mammal-like reptiles". The size of a badger, it had powerful jaws and mammal-like teeth.*

Alistair Barnard

formed a new jaw hinge directly with the squamosal bone of the skull. The mammalian postdentary bones and quadrate have lost their contact with the jaw altogether and have become the sound-conducting ossicles of the middle ear.

Cynodonts also evolved a much more mammalian kind of skeleton, with very slender limbs held much closer to the body, which made for more agile locomotion. It is often thought, too, that the cynodonts had evolved many of the physiological characters of mammals, such as full warm-bloodedness, although not everybody accepts this. There is no good evidence either way about whether they were furry.

Sometime around the end of the Triassic period, about 195 million years ago, the first fossils that have a fully developed new jaw hinge between the dentary and the skull are found. They are generally accepted as the earliest true mammals, and undoubtedly they evolved from advanced cynodonts. However, as they possess small postdentary bones and quadrate still attached to the lower jaw, they are very primitive. The best known of these earliest of mammals are the morganucodontids, which have been found in Europe, South Africa, North America, and China, which indicates that they had a worldwide distribution. They were all very small, with a skull length of 2 to 3 centimeters (just under to just over an inch) and an overall body length of around

Alistair Barnard

▲ *Morganucodontids, among the earliest of the true mammals, evolved about 195 million years ago. They were very small — about the size of a shrew.*

12 centimeters (about 5 inches). Their teeth were sharp and multi-cusped, and, from the wear patterns that developed during life, it seems that they were used to capture and chew insects and perhaps other terrestrial invertebrate prey. To judge from this and from their apparently

enhanced sense of hearing and smell, morganucodontids are believed to have been adapted to a nocturnal hunting existence.

THE MESOZOIC MAMMALS

Morganucodontids and related early mammals were the start of an evolutionary radiation, through the rest of the Mesozoic era, into several different groups. However, without exception they remained very small animals. Even the largest were no larger than a domestic cat, and the great majority were very much smaller than that. Furthermore, most of them remained insectivorous, although one group, the multituberculates, adopted a herbivorous habit and looked superficially like the rodents of today.

The most important evolutionary step during the Mesozoic was the development of a more complex kind of tooth, the tribosphenic molar. These teeth have a triangle of three main cusps on both the upper and the lower teeth, while the lower also have an extended basin at the back end. They were more effective in their action, and in the Late Cretaceous, around 80 million years ago, two main groups of fossil tribosphenic mammals can already be distinguished. These are the marsupials and the placentals, which were soon to be the dominant mammals of the world. Indeed, even before the end of the Cretaceous several different kinds of placentals had already evolved, and thus the scene was set for the great post-dinosaur radiation of mammals.

Why all the Mesozoic mammals should have remained so small is a mystery. Possibly it was simply because they could not compete with the dinosaurs in the habitats appropriate for large terrestrial animals and therefore had to remain restricted to a nocturnal way of life suitable for small animals. Alternatively, the climate may have been such that the physiology of these early mammals was unsuitable for large animals.

There is also still much argument about what caused the extinction of the dinosaurs and the numerous other groups of animals, and to a lesser extent plants, that suffered at this time. But what is clear is that the mammals not only survived, but were for some reason particularly well placed to take advantage of the relatively empty new world set before them.

THE TERTIARY RADIATIONS

The evolutionary story of mammals over the last 65 million years to the present day, that is the Tertiary era, began from the few groups of small, insect-eating mammals that survived the extinction of the end of the Cretaceous. The story is very complicated, because as well as the extinction of the dinosaurs, the end of the Cretaceous also saw the beginning of the break-up of the great land mass into separate pieces that proceeded to drift apart, eventually forming the continents familiar

Alistair Barnard

today. For a brief time, Australia and South America were connected to one another by Antarctica, but they were shortly to become completely isolated island continents, on which quite different groups of mammals evolved. On the other hand, North America, Europe, and Asia remained effectively connected to one another. A further complicating factor was a series of climatic changes that affected mammal evolution from time to time by causing the extinction of certain groups and allowing other, usually more advanced, groups to survive and flourish. Thus there was an almost continual turnover of species and families as the millions of years passed.

In any event, within no more than 5 million years or so, several important new kinds of mammals had evolved. Looking at the major part of the world, as represented by North America and Eurasia, the early new mammals to evolve from the little insectivorous placental ancestors included the creodonts, which were the first of the large predaceous mammals. These were very like modern carnivores, with specialized shearing molar teeth and powerful sharp-clawed limbs. Several other groups of mammals had evolved into

large, browsing herbivores, with various patterns of large, flattened grinding molars, and elongated legs with little hoofs on the ends of the toes to improve their running ability. Some were quite strange looking, such as the uintatheres of North America, which had bony protuberances over the skull, and *Arsinoitherium* from Egypt with its pair of massive horns. Small herbivores, occupying the niches later adopted by the rodents, were represented, surprisingly, by a group of primitive primates called the plesiadapids.

As time passed, new and more modern groups of animals evolved, although often in rather primitive and unfamiliar form. By about 50 million years ago the true Carnivora, and the earliest of the horse family represented by *Hyracotherium,* the eohippus, had appeared, as had the first whales (though these still had serrated teeth) and other groups later to be highly successful, such as the bats, rodents, artiodactyls, elephants, and lemurs.

Around 40 million years ago, the climate deteriorated for a while, causing the extinction of about one-third of the existing mammalian families. The main sufferers were the more archaic, primitive kinds, and consequently the fauna took

▲ Uintatherium *was one of the first large herbivores, comparable in size to a present-day African rhinoceros. It had three pairs of bony protruberances on its head.*

Alistair Barnard

▲ Indricotherium, a hornless form of rhinoceros that lived during the Miocene, attained a height of 5.5 meters (18 feet) at the shoulder and weighed perhaps 20 tonnes (over 40,000 pounds), making it the largest land mammal ever.

on an increasingly modern appearance. This phase was followed by a long period of very favorable climate referred to as the Miocene. It was the heyday of mammalian evolution, when more groups of mammals were present than ever before or since.

One of the most significant new developments of the Miocene was the spread of grasses, to form great plains on which herds of grazing animals could thrive. Horses, and even more so artiodactyls — deers, antelopes and so on — radiated. Although the majority of groups of mammals were those familiar today, they included many bizarre species, particularly among the perissodactyls, such as the chalicotheres, which had longer front than back legs, and *Indricotherium*, the largest land mammal ever to have lived, which stood over 5 meters at its shoulder. Among the carnivores were a number of large sabre-toothed cats.

During the Miocene the apes also evolved from more primitive primates, and a little later, about 4 million years ago, the first members of *Australopithecus* had appeared in Africa, as the immediate forerunners of the humans that were to arise 2 million years later.

SOUTH AMERICAN MAMMALS

During most of the Tertiary, South America was an isolated island continent, and consequently the history of its mammal fauna differed from that of the rest of the world. Several unique orders of

herbivorous placental mammals evolved, such as the litopterns which resembled horses but were entirely unrelated to them, and other orders superficially like rhinoceroses, hippopotamuses, and so on. Another important group were the edentates, with such strange animals as the huge armored *Glyptodon*, and the giant ground sloth *Megatherium*. This order still survives today and includes some of the oddest mammals, the armadillos and South American anteaters.

The most unexpected feature of South American mammal evolution concerns carnivores, for these consisted exclusively of marsupial mammals, a group absent from the Tertiary of North America, Europe, and Asia. There were smaller forms, such as the didelphids and caenolestids, which have survived to the present. But there were also large predators, the borhyaenids, which played the same role as carnivores in other parts of the world, with forms as large as lions, and even an equivalent to the placental sabre-toothed tiger in the form of *Thylacosmilus*, which had huge upper canine teeth.

The only placental mammals (apart from bats, which easily migrate worldwide) that managed to invade South America during the earlier Tertiary were rodents and primates. They probably came from North America via the Caribbean Islands, and once established they evolved into the characteristic South American groups, the hystricomorph rodents and the platyrrhine or New World monkeys.

Owing to tectonic changes about 3 million years ago, South America eventually drifted northwards to connect with the North American continent. Because of the ensuing climatic changes, there followed what is often called "the Great American Interchange", which resulted in a dramatic alteration to the South American mammalian fauna. The South American herbivore groups disappeared, to be replaced by invaders from the north: representatives of such modern placental groups as horses, tapirs, llamas, and deer. Similarly, the large marsupial carnivores disappeared and in their place came dogs, cats, bears, and so on. On the other hand, many remnants of the old South American fauna do persist, such as edentates, small marsupials, rodents, and monkeys; indeed several of them invaded North America, like the didelphid opossums and the porcupine.

It is particularly clear in South America how the present mammalian fauna is a result of many processes, evolutionary, biogeographic, and climatic, that together caused a long history of change.

AUSTRALIAN MAMMALS

The early history of Australian mammals is hardly known yet because there are few fossil deposits containing mammals until well into the Tertiary. Apart from a single find of the tooth of a primitive and early herbivore, there is no evidence that placental mammals reached Australia until late in the Tertiary, and then only as a few rodent and bat species from Asia. All other placental species, such as the dingo and rabbit, are human introductions.

The egg-laying monotremes diverged quite early from the tribosphenic mammals, the marsupials and placentals. Fossil teeth of a primitive monotreme have been discovered in the Cretaceous of Australia, but they never seem to have been a particularly important group.

With these minor exceptions, all Australian mammals, past as well as present, are marsupials. They radiated into many of the niches associated with placental mammals elsewhere in the world, such as the large carnivorous thylacines, the grazing, herd-living kangaroos, and the insectivore-like opossums.

THE PLEISTOCENE EXTINCTIONS

The final drama in the history of mammals occurred only 10,000 years ago, in the late Pleistocene epoch. A phase of extinction is shown by the disappearance from the fossil record of many mammal species, but mainly the largest forms. There had been giant members of many groups throughout the world, such as the familiar mammoths, Irish elk, and the sabre-toothed tigers. Less well known but equally remarkable were such animals as giant apes and wart hogs in Africa, a giant lemur in Madagascar, a giant tapir and ground sloth in South America, and giant kangaroo, wombat, and platypus in Australia. All these became extinct at about the same time.

It is much argued whether this was due to some climatic change that made life more difficult for larger species, or whether it was connected with the rapid spread of humans throughout the world at that time. Perhaps this was the first baleful effect of human beings upon the mammals of Earth.

TOM KEMP

▼ Glyptodon, closely related to armadillos, was roughly the size of a small motor car and was covered in a rigid bony carapace.

HABITATS & ADAPTATION

Mammals have evolved to successfully occupy nearly every habitat on the face of the Earth. Only the interior of the Antarctic continent remains uninhabited by mammals. To achieve their remarkable success, mammals have evolved various locomotor and feeding adaptations that permit them to survive and prosper in habitats as diverse as tropical rainforests, grasslands, tundra, and deserts. Much of the diversity in the external form of mammals reflects their adaptations to specific habitats or environments.

MAMMALS OF THE FOREST

Because trees are a key habitat component of forest-dwelling mammals, particularly as sources of food, shelter, and routes of travel and escape, many mammals of the forest are excellent climbers and exhibit various degrees of specialization for moving through the trees. In more generalized species, such as many mice and rats, adaptation takes the form of long tails, which act as counterbalances as these mammals move along branches. Other more highly specialized arboreal (tree-dwelling) species have prehensile tails that serve as a "fifth hand". Prehensile tails have evolved independently in forest-dwelling representatives of several groups of mammals, including New World opossums, Australian possums and cuscuses, South American anteaters, pangolins or scaly anteaters in Africa and southern Asia, New World monkeys, South American porcupines, and some tropical rats. Opposable big toes, or in some cases thumbs, which aid in grasping branches, are also specializations for movement through trees and have evolved in several groups of primates, New World opossums, Australian possums and cuscuses, the koala, and the noolbenger or honey possum. Gliding as a means of movement in forests has evolved independently in flying squirrels of North America and Eurasia, African scaly-tailed squirrels, colugos or gliding lemurs in the jungles of Southeast Asia, and in members of the Australian gliding marsupials.

Perhaps the most specialized arboreal mammals are sloths, which spend virtually their entire lives in trees, usually coming to the ground only to

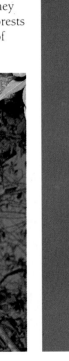

▼ *Most agile of arboreal monkeys of tropical America, spider monkeys swing through the trees with their long arms and legs and extremely prehensile tail: they are "five-limbed". However, the equally agile gibbons of Asia have no tail.*

Francois Gohier/Auscape International

defecate, on average about once every seven days.
Although almost helpless on the ground, sloths are
excellent swimmers, an adaptation to crossing the
numerous streams and rivers that characterize their
lowland forest habitats in the New World tropics.
The minimal level of activity so characteristic of
sloths is an adaptation to a diet of leaves containing
high concentrations of toxic compounds that are
manufactured by plants as a defense against
herbivores. The high fiber and cellulose content of
these leaves further reduces their nutritional value.
By coupling minimal activity with a low metabolic
rate and a dense coat for a tropical mammal, sloths
substantially reduce their food requirements and
consequently their intake of toxic chemicals. In
addition, sloths meet some of their energy
requirements by basking in the sun in the canopy
of trees. The koala, which also subsists on a diet
high in secondary compounds, possesses many of
the same adaptations, including slow deliberate
movements and relatively low metabolic rate.

Partridge Productions Ltd/Oxford Scientific Films

▲ The sloths of tropical and subtropical
American forests are adapted to a slow
way of life. Subsisting entirely on leaves
(which are a poor source of food), they
have a low metabolic rate, poor
temperature control, and very slow
movements. These strategies enable
them to be the dominant herbivores in
their environment.

◄ Many arboreal mammals leap from
one tree to another with legs
outstretched. It was a simple
evolutionary step to change a leap into
a glide by developing a web of skin
between the body and the legs — as in
this yellow-bellied glider, an Australian
marsupial. Gliding has evolved
independently in three marsupial
families, two rodent families, and the
flying lemurs or colugos.

Jean-Paul Ferrero/Auscape International

Richard Matthews/Planet Earth Pictures

▲ *The maned wolf has extremely long legs, employed in a bounding gait when pursuing prey — usually through tall grass. It is only distantly related to the true wolves and does not associate in packs.*

▼ *Ground squirrels are a major source of food for many carnivorous mammals, birds, and snakes. Out of their dens, they live in a constant state of nervous alertness. They benefit from living in social groups that provide multiple sense organs to detect predators.*

Francois Gohier/Auscape International

GRASSLAND MAMMALS

In contrast to forest-dwelling species, mammals living in grasslands often rely on speed, rather than concealment, as a method of escape. The long legs of hoofed mammals of the open plains function to increase their speed as a principal means of predator avoidance, as well as to permit them to efficiently travel long distances in search of food and water. One of the most striking grassland mammals is the maned wolf, a South American carnivore that has evolved extremely long legs to permit it to see over and move with ease through the tall grasses of the pampas.

The open nature of grasslands has promoted the evolution of social behavior in many species of grassland mammals, such as large ungulates (hoofed mammals), kangaroos, wallabies, ground squirrels, prairie dogs, baboons, and banded mongooses. Individuals in social groups benefit from multiple sense organs (ears, eyes, and noses) which aid in the detection of potential predators at greater distances than would be possible alone. Herding behavior in grassland ungulates is strongly reinforced by predators who selectively prey on individuals that do not remain part of the herd. Social predators such as lions and cheetahs are also characteristic of grassland habitats.

Many smaller grassland mammals are adapted for burrowing and utilize their burrow systems to escape predators and to provide relief from high temperatures during mid-summer days. Numerous members of the squirrel family, such as North American ground squirrels and prairie dogs, are semi-fossorial, meaning that they utilize extensive burrow systems for shelter and nest sites, but forage on the surface for herbaceous plants and seeds. Several groups of rodents, such as New World pocket gophers, South American tuco-tucos, and African mole-rats, have abandoned life on the surface to exploit rich subterranean plant resources in the form of roots, bulbs, and tubers. Pocket gophers and tuco-tucos utilize powerful forelimbs in burrowing, whereas African mole-rats use their large incisors to excavate burrows. These burrowing or fossorial rodents are characterized by short coats, small ears, small eyes, and short tails.

MAMMALS OF THE TUNDRA

Mammals of the Arctic tundra possess numerous adaptations to permit them to survive the physical rigors of this region. The immense shaggy coats of musk oxen mark this species as a true Ice Age survivor and one of the best adapted of Arctic mammals. One of the most familiar adaptations of Arctic mammals is the ability of some species to turn white in winter. The white winter coat of Arctic hares, Arctic foxes, and collared lemmings provides both camouflage in a white world and excellent insulation.

Locomotion during Arctic winters is often difficult, and tundra mammals possess several distinctive adaptations to increase efficiency. The surface area of the feet of Arctic hares is greatly increased by growth of dense brush-like fur on the pads. The unusually broad hooves of caribou aid locomotion in both winter snow and the marshy ground of summer. As an aid to digging through snow, ice, and frozen ground, the claws of the third and fourth toes on the forefeet of collared lemmings become greatly enlarged during winter.

Caribou or reindeer are unique among members of the deer family in that females typically possess antlers, albeit smaller than those of males. During winter both males and females use their antlers to scrape away snow and ice to expose vegetation on which they feed. Like many large mammals inhabiting grasslands, musk oxen and caribou have responded to living in an open environment by evolving herding behavior as a defense against predation. This is epitomized by the classic "circle the wagons" defense of musk oxen.

Another anti-predator adaptation in caribou is synchronized calving: 80 to 90 percent of calves in a population are born within a 10-day period in late May or early June each year. This brief calving season apparently represents a strategy designed to overwhelm and satiate predators with large numbers of young, and thereby allow some of the young to escape predation. The highly precocial newborn calves are soon able to join the herd and gain its protection. Although synchronized calving in caribou might represent a response to constraints imposed by the very short summer season of the high Arctic, synchronized calving also occurs in some grassland ungulates, such as the wildebeest in Africa, which are not under constraints of an extremely short growing season.

MAMMALS OF THE DESERT

Desert mammals must be adapted to succeed in harsh physical environments characterized by temperature extremes and shortages of moisture. Most desert mammals minimize the physical stress of these two factors by being active at night (nocturnal) and spending their days in burrows where temperatures are cooler and relative humidities are higher than on the surface. Insects and seeds are extensively utilized by desert mammals both for their nutritional value and as sources of water; the bodies of insects have a high moisture content. The moisture requirements of seed-eating rodents are met largely or entirely from water produced as a byproduct of their metabolism

summer

winter

▲ The third and fourth claws on the forefeet of the collared lemming become greatly enlarged each winter, as an aid to digging through frozen ground.

▼ A thick, shaggy coat protects the musk ox from the Subarctic cold, and adults are so large that they have no natural enemies. The young, however, are vulnerable to wolves. In the event of an attack, the adults surround the calves and form a defensive ring, with horns directed outwards: the classic "circle the wagons" defense.

Steve Kaufman/Bruce Coleman Ltd

▲ *The Arabian oryx inhabits deserts where the days are extremely hot and the nights correspondingly cold. In adaptation to this harsh environment, it feeds and moves mostly at night; its white coat reflects solar radiation; and its splayed feet are suited to walking on sand. Perhaps as a means of conserving energy, it does not maintain a constant body temperature, nor is there much aggression between members of a herd.*

of carbohydrates in seeds, termed metabolic water.

Antelope ground squirrels in the deserts of North America are active during the day and thus must directly confront the challenge of high environmental temperatures. Because lethal body temperatures are quickly reached while foraging, these squirrels periodically return to their burrows to "dump" heat in the cooler environment of the burrow. They resume their foraging after the body temperature returns to normal.

Large desert mammals, such as camels and various species of oryx, cannot escape to burrows and are exposed to daily temperature extremes. One way they survive without expending excessive amounts of water on evaporative cooling is to let their body temperature fluctuate. At night they take advantage of radiational cooling to allow the body temperature to fall below what is normal for mammals. During the day, they permit the body temperature to slowly rise as they are warmed by the sun, and only late in the day as the body temperature approaches lethal limits do they utilize evaporative cooling, thus conserving considerable water. Their pale coat helps both to reflect sunlight and to provide insulation against high environmental temperatures. Oryx also feed at night when the moisture content of plants is higher than during the day.

CONVERGENT EVOLUTION

In examining the adaptations of mammals to various habitats, it is obvious that certain adaptive types appear best suited to particular habitats. As a consequence, we find many examples of unrelated mammals that have evolved remarkably similar adaptations because they live in similar habitats and have similar ecological roles in different parts of the world. This phenomenon is termed convergent evolution.

Among desert mammals a notable example of convergent evolution involves seed-eating rodents, including kangaroo rats in North America, gerbils and jirds in Asia and Africa, jerboas in Asia and Africa, and Australian hopping mice. These rodents exhibit remarkable degrees of convergence in morphology and ecology. All have pale, sand-colored coats, enlarged hindlegs and long tails, and employ bipedal or hopping locomotion as a principal means of escape from predators.

The convergence of seed-eating desert rodents as well as other examples noted in this chapter suggests that there are optimal evolutionary answers to the challenges of adapting to exploit specific types of habitats. Because of their widespread distribution and their adaptability, mammals are excellent examples of convergent evolution.

GORDON L. KIRKLAND, JR

MAMMAL BEHAVIOR

All animals behave—that is, they react to specific stimuli in specific ways—but mammalian behavior is characterized by the fact that mammals, with only two exceptions, give birth to living young. In addition, the young of all species are dependent upon their mothers for food, in the form of milk secreted by specialized and modified sweat glands. This necessitates a special type of social behavior that bonds the mother to its young.

DEPENDENT MAMMALS

Some types of mammals remain dependent upon their mothers for many months—even years; such is the case with primates (including humans), and with whales and elephants. Others, such as rodents, are weaned in less than three weeks. Whatever the length of this dependence, it is an example of social behavior that can be defined as an interaction between two individuals of the same species. Even the so-called "solitary" species of mammal, such as the Australian bandicoot or the orang utan, which as adults interact socially only to mate, have a close behavioral interaction with their mothers when they are young. All other behaviors shown by mammals are classified as something other than social. These include behaviors necessary for bodily maintenance, such as feeding, grooming, urinating, defecating, and huddling.

MAINTENANCE BEHAVIOR

It is difficult to dissociate maintenance behavior completely from social behaviors, since frequently the two are interrelated. For example, passing urine is a biological necessity associated with the physiological processes of nitrogen excretion. Cows, seals, whales, and female dogs and mice simply void urine whenever the distension of the bladder reaches the point where the urination center in the rear of the brain is stimulated. After urination, the stretch receptors in the bladder wall relax, the flow of messages to the brain stops, and the animal no longer feels the urge to urinate.

Adult male dogs, mice, and many other mammals, however, do not urinate in this way. In them the presence of the sex hormone testosterone—produced in the testes—influences behavior, so that urine is voided at specific points around the animal's environment where other males will encounter it. This is an example of territorial demarcation. So maintenance and social behaviors are sometimes intertwined.

The same is true for feeding behavior—many carnivorous animals hunt cooperatively in order to fill their bellies—and grooming behaviors which,

▼ Food supply is the responsibility of the female members of a pride of lions. Individual females may capture small prey, but two or more usually cooperate in bringing down antelopes, gazelles, and zebras. Even with such relatively large prey, cost-benefit considerations apply: if the animal cannot be caught after a relatively short dash, it is not worth expending more energy in a prolonged chase.

Purdy & Matthews/Planet Earth Pictures

in primates, serve an important social function. Nevertheless, in all these behaviors the underlying drive is the maintenance of the individual; the social aspect is superimposed upon it.

SOCIAL BEHAVIOR

Apart from the relatively short interactions between mothers and their young, social behavior may be divided into four main categories: play, the acquisition of social dominance, the acquisition of territory, and sexual behavior. The eventual goal of all of these behaviors is to differentiate the fitter individuals from the less fit, allowing only the fittest to mate and to pass on their genes. The notion of equality has no place in nature.

Play

Play is a special form of social behavior indulged in by the young of primates and of the carnivorous mammals, especially dogs, cats, weasels, and mongooses. It is noticeably absent in the young of the large herbivores, although occasional glimpses may be seen in lambs, foals, and calves. Play serves to provide the young animal with an opportunity to develop the skills it will later need to support itself in the wild. Young carnivores invest much time in mock fights, during which they practise attacking. Watch two puppies tumble about: they make frequent lunges to the throat and neck area, where a killing bite will be most effective. Primates are the most highly social of all mammals, and among them play serves to teach the young how to form social relationships and how to relate to other individuals. Play also teaches them how to react to signals sent out by another individual which indicate social status and mood.

▼ Young chamois at play. Play is usually restricted to the young of social species while they are under the protection of adults. It involves "sketching" or "rehearsal" of behavior — particularly food capture and combat — that will be important in later life. Play also improves an individual's agility, a matter of importance in the rock-climbing chamois.

Gunter Ziesler/Bruce Coleman Ltd

Acquiring social dominance

Many mammals do not live in one place all the time, but migrate from one place to another. Frequently they manifest a social hierarchy, in which one—or a few—individuals are dominant over all the others. In a troop of baboons, for example, the dominant males travel in the center, while younger aspiring males make up the vanguard and rearguard. The dominant males are easily distinguishable by their size and by the thickness of their manes. Some sort of adornment makes dominant members of any group distinctive: it may be visual, as in the baboon or the blackbuck, in which only the dominant buck is black (all other males are light brown in color); acoustic, as in the South American howler monkey, in which the dominant males have the loudest voices; or olfactory, as in the Australian sugar glider, in which the scent gland on the heads of the most dominant males secrete more copiously than

Gerald Cubitt/Bruce Coleman Ltd

◄ In most social mammals, one individual is the leader of a group. In some species the dominant individual is a female, but much more commonly it is a male that has attained the position after competition with rivals and monopolizes (or attempts to monopolize) breeding. This herd of blackbuck is led by the strikingly marked male; the dark individual in the background is a subadult male, too young to be a threat to the leader.

▼ Individuals (or pairs) of many species occupy well-defined territories and attack trespassers — which usually retreat. When two individuals from neighboring territories meet at a border, the tendencies of each to attack or retreat are delicately balanced, leading to a "stand off", as in these short-tailed shrews.

those of the subordinate males. In all cases, the dominants are responsible for most of the matings.

Acquiring territory

A large number of mammals are territorial, which means that the males compete for, and repel all other males from, a piece of land. The territory may be very large, as is the case with the large cats, containing enough food and other resources for rearing a family, or it may be a token patch barely big enough for the male and his consorts. The latter type of territory is seen in elephant seals and some ungulates, such as the Uganda kob antelope. Territorial fights may be aggressive, but as a rule the loser is not killed. Horns and antlers have evolved to minimize the risk of accidental death through goring. In most cases the territorial sex is the male, who may be substantially larger than the females, in order to give him the best chance of winning his encounter. Almost invariably the

Dwight R. Kuhn

males show the visual, acoustic, or olfactory adornment necessary to demarcate and defend the territory against infiltration. Among small and medium-sized mammals, scent marking of obvious points on the boundary and within the hinterland of territory is commonplace. Such scenting places, redolent of the scent signatures of their visitors, act as information exchanges for all who pass by.

Sexual behavior
Mammalian sexual behavior varies from the short and rough, as seen in bandicoots, to the highly stylized and ritualized performances of creatures such as the Uganda kob antelope. Whether mating is promiscuous, as in these two examples, or whether male and female mate for life, as in many of the small forest-dwelling antelopes, or whether it is something in between, as in most mammals, depends on many complex and intertwined environmental factors. Non-promiscuous species show a degree of courtship, during which a male and female annexe themselves from the group. In a

number of species, the close proximity of the male acts to bring the female on heat, and only at this time will he attempt to mate. It is thought that pheromones, or signalling odors, in the male's urine act to stimulate the female's reproductive system. In some species—the domestic dog and the hamster are familiar examples—the female may emit a special odor which lures males from afar. She may then choose with whom she will mate. In those species in which both sexes need to help with rearing the young, for instance most primates and many carnivores, courtship will be of sufficient duration for the offspring of any previous alliance to become apparent before mating occurs. If the female is pregnant, a pair-bond may not form. In this way the male will be assured that the subsequent offspring, which he is helping to rear, is truly his own. Courtship in these species is characterized by subtle and covert cues, not intended for a general audience.

In stark contrast, those species that mate promiscuously advertise their sexual readiness in

▼ *When not resting, mature male lions are occupied with defending a territory against other males and with mating. Mating involves rather fine judgement on the part of the male, since a female that is not on heat will reject his advances quite vigorously.*

D. Parer & E. Parer-Cook/Auscape International

Jonathan Scott/Planet Earth Pictures

▲ Mutual grooming in olive baboons. Reciprocal grooming is common in mammals and birds, usually serving to reinforce a pair-bond. In monkeys and apes, mutual grooming frequently extends throughout a group as an expression of group solidarity and of subtle differences in social rank.

flamboyant behaviors. Female chimpanzees, who are strongly promiscuous, are adorned with a large patch of white sexual skin in the perivaginal area which inflates massively during the heat. Males are strongly attracted to this and line up patiently waiting their turn in the mating queue. Interestingly there is little interference and aggression seen in this whole bizarre process.

THE SURVIVAL OF THE SPECIES

Mammalian behavior is not only complex, but also highly adaptive—that is, it contributes to the survival of the species just as much as any other biological attribute. Research into mammalian groups that occupy a wide range of habitats, such as primates or antelopes, is revealing that the ultimate factor that governs which type or behavior evolves is the environment. Thus the strange behavior of marmosets, in which the female always has twins but is large enough to carry only one youngster, may be explained by considering their territories. These are so large in relation to the tiny size of the animals that they can only be adequately defended by both the male and the female—a female on her own is unable to rear a family and protect the territory. Natural selection has evolved a reproductive strategy in which, if the male does not remain with the female to help carry and protect the young—and incidentally help defend the territory—his genetic investment in the next generation will be impaired. This apparent act of genetic blackmail on the part of the female is no more than the influence of the environment on aspects of the species biology that lead to increased chances of survival.

D. MICHAEL STODDART

C.A. Henley/Auscape International

▲ Female brown antechinus with 8-week-old young. At birth these marsupials, not much larger than a grain of wheat, attach themselves to the mother's teats, where they remain for about five weeks. They are then left in a nest and suckled daily for another eight weeks.

Frieder Sauer/Bruce Coleman Ltd

◄ Male impalas rubbing heads. For much of the year, male and female impalas associate in separate herds. Within a male herd, individuals may signal group ties by rubbing heads and exchanging scents produced by glands on their faces. These relationships break down in the breeding season.

ENDANGERED SPECIES

Ever since the first living forms inhabited the Earth over 3.5 thousand million years ago, species have become extinct. During that huge span of time, an estimated 250 to 500 million different forms of life have existed, and 98 percent of these have gradually died out and have been replaced with new types. The average worldwide rate of extinction of all species over that time was somewhere between a minimum of one species per thousand years and a maximum of one species per year. But extinction is speeding up. Today, everywhere on Earth, species are disappearing at an ever faster rate—be they birds, fishes, plants, mammals, invertebrates, or reptiles. And they are not being replaced.

"EXTINCTION SPASM"

The life forms existing today number between 3 and 10 million species, half of which live in tropical rainforests. No one knows exactly how many there are, because possibly millions—mainly invertebrates such as insects and lower plants—have yet to be discovered and named. Zoologists are certain, however, that approximately 4,300 kinds of mammals exist. That's all.

Between 1600 and 1900, scientists estimate that 75 species, mostly birds and mammals, were killed off. Another 75 disappeared between 1900 and the 1960s. Since then, the rate has soared. Dr Norman Myers, the British ecologist, warned that we can now expect to lose anywhere from a minimum of 1,000 species of plants and animals per year to a maximum of 100 species of plants and animals per day. During the 1990s, the number of mammal species considered threatened more than doubled. In 1996 the IUCN listed 1,096 species, a quarter of all the known mammal species, as vulnerable or endangered; an increase from 535 in 1990 and 741 in 1994.

If this alarming trend continues then virtually all mammals will be endangered by 2008. By 2015, readers still alive will have witnessed the extinction of maybe a million living things. Such a violent and rapid event has never before happened in Earth's long history. It is called an "extinction spasm".

THE HUMAN FACTOR

What is causing this crisis, and how did it begin?

▼ *The black-footed ferret became extinct in the wild when human activities eliminated its major food source, prairie dogs, from its habitat. Two United States government agencies are breeding the last remaining black-footed ferrets in captivity in the hope of reestablishing this animal in its original habitat.*

Gerard Loez/NHPA

The human population explosion is probably a significant factor. Human beings are exploiters. We always have been; we always will be. Humans are among the smartest and most skillful of mammals. It hasn't taken us long, in geological time, to gain control over other mammals. Ever since we emerged as intelligent hunters and food-gatherers, we have deliberately utilized other mammals for food, milk, clothing, footwear, weapons, tools, and oil. And over the last few thousand years we have successfully spread into every nook and cranny of the world. With our highly organized societies, our increasingly sophisticated forms of transportation, weaponry, and communication, and our ever more demanding types of agriculture and mariculture, we rule the planet. At the same time, human numbers have skyrocketed. We are now the major threat to most other life forms and the leading cause of biotic impoverishment.

A key factor has been the way crude weapons have evolved into efficient and fearsome killing machines. Once hunters used clubs and rocks, then sling-shots and spears. They moved on to bows and arrows, and then firearms, progressing from single-shot muskets to explosive harpoon guns to rapid-fire machine-guns. Transportation

has developed in a similar way—from using two feet, to riding beasts of burden, to piloting sailboats and wagons, to running trains, planes, ships, and cars with fossil fuels.

The dreadful impact of these developments on wildlife can be shown by case histories. For example, archeologists sleuthing through ancient "boneyards" of the late Pleistocene (100,000 to 10,000 years ago) have found proof that prehistoric people, despite their small size and puny weapons, were able to group together to kill the huge mammals, or megafauna, of that era in large numbers. Skeletons have been unearthed with stone arrowheads or spearheads in them.

Each time that humans arrived on a new land mass, a wave of extinctions appeared to follow. For instance, shortly after humans crossed the Bering Strait to North America, around 10,000 to 15,000 years ago, toward the end of the Ice Age, there was a swift disappearance of giant beavers, elephants, mastodons, camels, wooly mammoths, sabre-toothed tigers, and giant bison. At least 50 types of megafauna vanished. The same coincidental extinctions seem to have occurred in Europe, Africa, Latin America, and Australasia.

The Malagasy Republic (Madagascar) is an area

▲ *Wild tiger numbers hover at around 6,000. Since it is impossible for humans and tigers to live in harmony, the species can only be conserved by setting aside areas of forest large enough to provide prey for these carnivores. By conserving tigers, we simultaneously conserve thousands of other species of plants and animals.*

Richard Matthews/Planet Earth Pictures

▲ The golden lion tamarin of South America. Until recently, the most vulnerable forests in South America were the hardwood forests in the coastal region of southeastern Brazil. By the early 1970s, however, they were wiped out save for a few patches—and with them went their primates, including the golden lion tamarin. The three species were reduced to tiny remnant populations and today there are fewer than 600 individuals in the wild. Despite attempts to reestablish wild populations with captive-bred animals, these species remain endangered.

that was more recently invaded by human beings. The Indonesians arrived about AD 20, and the disappearance of giant lemurs, elephant birds, and dwarf hippos followed. A manmade overkill took place. No sign of climatic change or other biological threats exists to explain the die-offs.

Neither the Ice Age nor the Madagascar extinctions, however, seem as terrible as the slaughters during historical times. After white people arrived in America, they practised roundups and slayings of millions of bison and other large mammals. One such event was the Great Pennsylvania Circle Hunt. In 1760, hunters on foot formed a 170 kilometer (100 mile) circle. Spaced just under 1 kilometer (about half a mile) apart, they marched inward, killing everything they met with guns. The total for the hunt: 41 cougars, 109 wolves, 112 foxes, 1 otter, 12

wolverines, 3 beavers, 114 bobcats, 10 black bears, 2 elk, 98 deer, 111 bison, and 3 fishers!

More advanced weaponry and transportation allow humans to hunt wolves and polar bears today in Alaska. Hunters use light aircraft to track, tire, and approach wolves on land, while polar bears are scouted on the immense ice floes by helicopter. The animals are then shot on land using high-powered rifles with scopes. Such high-technology killing denies game animals any chance of survival.

WORLDWIDE HABITAT DESTRUCTION

Mammals today, however, are most at risk from the environmental problems caused by humans. Among the worst is the destruction of their habitat—every ecosystem on Earth is being tampered with as people dam rivers, log woodlands, drain wetlands, build pipelines, graze domestic stock, and construct highways and suburbs. Mammals are rapidly losing their homes, food, and water, and open space.

The largest single threat is clearing of land. Although the area of land in parks and reserves has doubled during the 1990s, the rate of land clearing has more than doubled. Development in the Amazon rainforest, which supports half the Earth's living species, is the most frightening example.

The Amazon basin is so vast—6 million square kilometers (over 2 million square miles)—so seemingly fertile, so green, that it seems like an untapped paradise. Several Latin American nations, which own part of it, are trying to colonize it. Brazil is the leading country. Its government built the lengthy Transamazonica Highway, and many spur roads, in order to resettle thousands of poor peasants along its edges. These colonists have each cut and burned small tracts of virgin forest to plant rice, pepper, bananas, and cacao. When colonist cropland is hacked out of the jungle, the habitat of jaguars, howler monkeys, scarlet macaws, and other rainforest animals fragments and diminishes.

Most of the Amazon basin is, in fact, a false paradise for colonists. Soils are infertile and contain toxic aluminum, and rains are torrential. Once the green umbrella of tall trees is gone, soils rapidly compact and erode under the scorching sun and harsh rains. Eventually, they are worthless. Most farming projects have failed.

Now the Brazilian government is promoting huge cattle ranches and timber or palm plantations. Most are owned by multinational corporations and cover thousands of hectares. When the rainforest is cleared, huge fires burn wildly for weeks. In fact, the largest manmade fire ever reported (detected by satellite) was set by a multinational corporation to prepare pastures.

It is estimated that an average of 20 hectares (50 acres) of rainforest is burned or disturbed in the world *each minute*. That's equivalent to an area the size of Great Britain *every year*. Already, 37 percent of the moist tropical forest in Latin America has been converted into less productive ecosystems; 38 percent in Southeast Asia; and 52 percent in Africa. This astonishing changeover from a complex, balanced, healthy ecosystem with huge mature forest trees, to short-term cropland with oil palms and pepper bushes, or land for cattle grazing, is one of the most disturbing ecological and economic trade-offs happening today. If the destruction continues, scientists predict that by about 2010 only pockets of rainforest will remain in remote places.

Another form of habitat destruction is occurring in northern and western Africa. A region called the Sahel is experiencing swift desertification. Largely because of overgrazing by domestic animals, poor rainfall, and high population rates, the Sahara Desert is stealthily encroaching on savanna lands. An estimated 250 million humans were recently stricken by drought, along with their herds, and wild mammals such as gazelles, antelope, cheetahs, and addaxes died of thirst.

THREATS FROM ALL SIDES

Many environmental abuses indirectly affect mammals. Such abuses include air pollution, such as acid rain, fresh and salt water pollution, and soil degradation. We have no way of knowing how many mammals take sick or die from drinking filthy water, eating plants growing in contaminated earth, or living in sea water polluted with toxic and nuclear wastes. Research carried out on roe deer living in a Polish national forest downwind of a major steel manufacturing city showed deformities in antler growth and a declining birth rate. Wildlife biologists have surmised that the deer eat vegetation that has been doused with acid rain and contains traces of toxic metals from the steel mills. Reindeer of Lapland are known to have become radioactive after eating lichens that absorbed nuclear fallout from the Chernobyl disaster in the Soviet Union.

One looming environmental factor is global warming. As more and more carbon dioxide and methane are pumped into the atmosphere from tropical wildfires and combustion of fossil fuels, the Earth will heat up. A Pandora's box of climatic changes lies in store: warmer temperatures, more intense storms, rising sea levels, and torrential rains at latitudes unaccustomed to them. Some mammals will be able to migrate to find comfortable new locales, and some may adapt. Others will become too stressed and will succumb. Global warming may produce the greatest extinction spasm ever.

The introduction of exotic animals into existing ecosystems, for food or recreational hunting, is another threat to native wildlife. Foreign species can not only bring in and spread unsuspected diseases or parasites, but they are also more likely to survive than residents and often experience a population explosion. Fierce competition ensues and local fauna can be destroyed.

All kinds of human activity—including scientific research, and the testing of various cosmetics and drugs—continue to destroy wildlife. Human aggression during wars kills countless mammals. So does exploitation for the production of fashion items.

THE TALE OF FIVE SPECIES

To illustrate the dangers confronting mammals, here are five examples from the land, air, and sea. All five species are listed as endangered in the Red Data Book of the International Union for the Conservation of Nature.

Maned wolf

On the dry, grassy pampas of Argentina, Paraguay, and Brazil roams the maned wolf. It looks like a rangy, red dog on stilts, with 18 centimeter (7 inch) ears, black stockings, and a black "cape" over the shoulders. Weighing over 20 kilograms (close to 50 pounds), it is wolf-sized but is actually a fox. Maned wolves can cover over 30 kilometers (20 miles) a night on their long legs, and hunt mainly for pacas, rodents, rabbits, and birds. They are hunted by collectors for zoos and by local people for charms: in Brazil, people believe that the body parts have medicinal value and that the left eye—if plucked from a *live* animal—is lucky. Because of these problems, plus parasites and diseases, fewer than 2,000 maned wolves are left today.

▼ *Slaughtered for the supposed magical properties of its parts, the maned wolf of South America is now reduced to about 2,000 animals and is unlikely to survive in the wild.*

Francisco Erize/Bruce Coleman Ltd

Coo-ee Historical Picture Library

▼▲ The existence of the American bison was not threatened until the arrival of Europeans. Between the middle and the end of the last century, "buffalo" hunters reduced the population from about 50 million to fewer than 800. Protection and management in reserves has permitted numbers to increase to about 40,000.

American bison

The American bison and its close relative the European bison (a subspecies) have both had near brushes with extinction in the past hundred or so years. Once, an estimated 50 to 60 million animals covered the American prairies, looking like an ocean of black dots. Migrating bison were considered the greatest natural spectacle that humans had ever seen. The Plains Indians depended heavily on bison, or buffalo as they became known, and revered them too. But it was the arrival of Europeans that almost eliminated the shaggy 1 tonne (2,200 pound) beasts.

Professional hide and meat hunters slaughtered bison by the millions. Often only the tongues, considered a delicacy, were taken and the meat left to rot. Buffalo Bill once counted 4,280 animals he alone killed in one year. Railways pushing west also played a part in bison destruction. Passengers on the early trains took pot shots at buffalo simply to while away the time. By 1884 the bison had practically disappeared from the United States, but at the last moment, reserves were established and the few remaining animals protected. There are now several thousand in Yellowstone and Wood Buffalo National Parks, the Wichita Mountains Wildlife Refuge, and in other refuges. But because the animals are confined to these isolated sanctuaries, there is little genetic exchange. Inbreeding and gene loss will possibly harm the species. The bison may be "saved", but huge herds will never again thunder across the plains.

Humpback whale

The fate of the humpback whale is a striking example of exploitation. The 19 meter (62 foot),

Jeff Foott/Bruce Coleman Ltd

James Watt/Planet Earth Pictures

black-colored baleen feeders once numbered 100,000. By the 1960s, only 6,000 remained.

People have hunted whales for centuries. Norse fishermen made whale carvings 4,000 years ago and Eskimos have eaten them for at least 3,500 years. But commercial whaling, the real culprit, started as early as the tenth to twelfth centuries AD. Whales were hunted for their meat, oil, and bones. Humpbacks were one of the easiest whales to catch, so they were harpooned by the hundreds. As long as whalers used small boats and killed and butchered by hand, whales had a chance. But when huge factory ships with sonar, helicopters, explosive harpoon guns, and onboard processing appeared, whales were doomed.

In 1946, the International Whaling Commission was established to regulate whaling. Some nations paid no attention. In 1973 the United States Marine Mammal Act prohibited taking these animals and their products, except by native peoples. At the same time, the United Nations Conference on the Human Environment set a worldwide ten year moratorium on whaling. The Commission did not abide by this ruling. Finally, thanks to the concern of many scientists and the

growing concern of the public, a ban on all commercial whaling took effect in 1986. Conservation groups like Greenpeace and their Save the Whales campaign have helped people to learn about and respect whales. Modern technology allows humans to hear the odd patterns of chirps, groans, and cries that male humpbacks sing in the breeding season. Female whales (and hydrophones) can hear these songs up to 160 kilometers (100 miles) away under water.

This is the first time in human history that whales have been protected. Even so, Japan, Iceland, and Norway still harvest whales under the guise of research, and 95 percent of the humpbacks are now gone.

Flying foxes

Bats make up nearly one-quarter of all mammal species. Of these, the 173 different kinds of flying fox are the most fascinating. The largest bat on Earth is the Samoan flying fox, which has a wing span of nearly 2 meters (6 feet). Since these bats feed on flowers and fruits, nectar and pollen, they use their keen eyes and noses rather than sharp hearing and echolocation to find food.

▲ The use of the harpoon gun and factory ships brought most of the large whales to the verge of extinction. Since a recent ban on commercial whaling, numbers of some species, such as the humpback whale shown here, appear to be increasing slowly.

These secretive creatures with large, luminous eyes and dog-like faces are hunted for meat in Malaysia and Indonesia, where their flesh is considered a delicacy. Bat meat is also thought to cure asthma. A hunter with a modern shotgun can exterminate a colony of a thousand flying foxes in a short time. Some bats are only wounded, however, or become orphaned youngsters. They crawl away to die and are never collected or used.

Many tropical plants rely on fruit bats for seed dispersal and pollination. These "bat plants" give humans more than 450 useful products, such as black dye for baskets, delicious fruits, kapok fiber, charcoal, and medicines. In West Africa, the straw-colored flying fox disperses seeds of the iroko tree, which is the basis of a $100-million-a-year industry.

But flying foxes are declining, even disappearing, on many islands, including Guam, the Mariana chain, Yap, Samoa, Rodrigues, and the Seychelles. The destruction of their habitats by agriculture and fuelwood gathering is the main threat. Little conservation work is done locally. But in Malaysia and Indonesia, the World Wide Fund for Nature has promoted public seminars and news coverage on the value of flying foxes and other bats.

Gorillas

Anyone who has read *National Geographic*'s article on gorillas, or the book *Gorillas in the Mist* by Dian Fossey, or seen the movie of the same name, knows the truth about these largest of living primates. They are, indeed, gentle giants.

Dr Fossey spent almost 20 years studying them in Africa and developed a close relationship with several groups of mountain gorillas. Despite her work and the breakthroughs she made, humans remain the gorillas' worst enemy. Even though African laws prohibit killing gorillas, enforcement is usually lacking. Poachers still sell the heads and hands as souvenirs. The lowland subspecies is now estimated to have a wild population of between 30,000 and 50,000, but the highland populations have been badly affected by civil war in Rwanda and there are now as few as 350 to 500 mountain gorillas left. Even in the Virunga Volcanoes National Park, the gorilla habitat is not safe. Recently, a sizable section was gobbled up to grow pyrethrum, an insecticide, and loggers are constantly nibbling their way up the volcano slopes.

The greatest hope of saving gorillas is through tourism. Already, special groups visit Virunga to photograph the primates without harming them in any way. This leads also to local residents, rather than poachers, gaining income.

CONSERVATION CAN STOP THE ROT

As grim as the outlook is for many mammals, conservation efforts are underway in almost every country. Hundreds of government agencies and non-government organizations exist. Among the notable are the International Union for the Conservation of Nature (IUCN), the World Wide Fund for Nature, and the United National Educational, Scientific, and Cultural Organization (UNESCO). In the United States, the National Audubon Society, the Sierra Club, the Nature Conservancy, and Planned Parenthood are also prominent.

Many laws protect wildlife and wildlands all over the world, but the problem lies in providing sufficient enforcement. International cooperation is, however, often very successful in saving species. For example, the population of the elephant, the largest land mammal, plummeted from 1.3 million to 625,000. Poachers and rebels

▼ Flying foxes are usually numerous, but their habit of congregating in dense roosts makes them vulnerable to mass slaughter. Island species such as the Marianas fruit bat have become endangered by human predation.

Merlin D. Tuttle/Bat Conservation International

D. Parer & E. Parer-Cook/Auscape International

still shoot elephants for ivory to provide income and finance civil wars, but since ivory has been on a list of prohibited products compiled by CITES (Convention on International Trade in Endangered Species), whose rules are observed by 75 nations, the animals have had a chance.

Captive breeding in zoos can help some endangered species. These establishments have bred 19 percent of all living mammal species over the years. A particularly successful captive breeding program has been that of the golden lion tamarin, and there has been some success with the Arabian oryx. Recently, some of the tamarins have been released back into small pockets of suitable habitat, but efforts to reintroduce species to the wild are as yet in their infancy. The main function of zoos remains the education of the public about habitat destruction and disappearing species.

Game ranching and farming is another technique for fostering rare animals. Elands, wildebeeste, wild pigs, capybaras, and pacas are some that can be raised profitably. Often, game meat contains more protein, less fat, and brings in more money than meat from domestic animals. In Venezuela, capybara meat may be eaten during Lent whereas all other animal meat

may not—Catholic monks in the sixteenth century considered them fish!

Conservation efforts to maintain populations of some of the more spectacular mammalian species are barely holding their own or making very slow progress. However, it is the smaller, less obvious, mammals like bats, shrews and mice which are most in peril. Being small, nocturnal, and inconspicuous may help protect these animals from natural enemies but it offers no protection against the bulldozer, the chainsaw, and the firebrand.

Without question, the preservation of habitat through outright purchase, lease, easements, or donations is the key to saving species. At present, national parks and equivalent reserves protect less than 2 percent of the world's land surface. It's not enough. Reserves big enough to protect all resident species in every ecosystem are urgently needed.

If we want to stop the slide towards extinction of the world's remaining mammals, we must act immediately to halt clearing of land. Ultimately, human attitudes must expand beyond utilitarian, economic, or recreational interests in animals and plants. All living things should be valued in and of themselves and have the right to exist.

ORIGINAL TEXT BY ANNE LABASTILLE;
REVISED BY GEORGE MCKAY

▲ Gorillas are threatened by reduction of habitat, poaching of young animals for some unscrupulous zoos, and slaughter for souvenirs. Controlled tourism may provide sufficient revenue to encourage local people to protect this species.

PART TWO
KINDS OF

MAMMALS

MONOTREMES

SIZE

Platypus *Ornithorhynchus anatinus*
Length: 39–63 cm (15–25 in)
Weight: 0.7–3 kg (1½–6½ lb)

Short-beaked echidna
Tachyglossus aculeatus
Length: 30–45 cm (12–18 in)
Weight: 2–7 kg (4–15 lb)

Long-beaked echidna *Zaglossus bruijnii*
Length: 60–100 cm (24–39 in)
Weight: 6–10 kg (13–22 lb)

CONSERVATION WATCH
!! Long-beaked echidna
populations are endangered.

▼ *The platypus propels itself through the water with its forefeet, using its hindfeet to control or change direction. The white patch of fur beneath the eye closes the eye and the ear opening when the animal is underwater.*

There are only three living species of monotreme in the world: the duck-billed platypus *Ornithorhynchus anatinus*, the long-beaked echidna *Zaglossus bruijnii*, and the short-beaked echidna *Tachyglossus aculeatus*. The platypus, found only in the eastern states of Australia, is such an unusual animal that when the first specimen was sent back to Britain in 1798 it was widely assumed to be a fake, made by stitching together the beak of a duck and the body parts of a mammal! The two species of echidna, or spiny anteater, are found on the island of New Guinea, and the short-beaked echidna is also widely distributed in Australia, where the long-beaked echidna is now unknown. Current fossil evidence suggests that this small group of mammals may have originated and diversified in the Australian–Antarctic section of Gondwanaland.

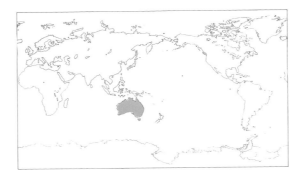

MAMMALS THAT LAY EGGS

The earliest Australian monotreme fossils so far discovered are from the early Cretaceous Period (about 100 million years ago). Although long-beaked echidnas are no longer found in Australia, fossil species did inhabit the Australian mainland and Tasmania until the late Pleistocene period (around 15,000 years ago).

The order name Monotremata, meaning "one hole", refers to the fact that all three species have only one opening to the outside of their bodies, through which the products of the reproductive, digestive, and excretory systems are passed. This feature is not exclusive to the monotremes, as marsupials also have this arrangement. While all three species have some anatomical similarities to reptiles—for example, they have extra bones in their pectoral, or shoulder, girdles—their anatomy and physiology are largely mammalian in nature.

But the most distinctive feature of the group is that they lay eggs, instead of giving birth to live young as all other mammalian species do. The eggs are soft-shelled, measure approximately 13 x 17 millimeters (½ x ⅔ inch) and are incubated for around 10 days, after which the young are fed milk. This suckling of the young, as well as the possession of hair, a single bone making up each side of the lower jaw, and a muscular diaphragm

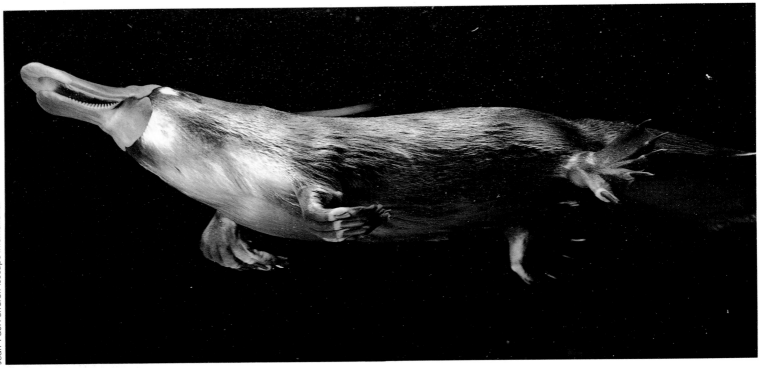

A UNIQUE BUT VENOMOUS AUSTRALIAN

The platypus is one of a very few venomous mammals. Two species of Caribbean solenodon and a few species of shrew use poisonous saliva to subdue prey, often larger than themselves, but the platypus has something completely different and of unknown function.

Rear-ankle spurs are found in all three species of monotreme. With their associated glands, they are known as the crural system. In females of all species of monotreme the spur is lost during the first year of life. Although the spurs and glands persist in male echidnas of both species, echidnas do not seem to use the system to inject venom; the platypus does.

Changes in the structure of the spur in male platypuses can be used to age animals up to 15 months after they have left the breeding burrow. Fully grown adult spurs are around 15 millimeters (½ inch) in length, can be averted away from the ankles, and can be driven into an object by the action of the muscles of the rear legs. The puncture alone is painful, but the venom injected can lead to symptoms ranging from local pain and swelling to paralysis of a whole limb in humans. When the species was

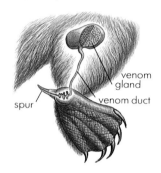

Curved and hollow, the platypus's spur (above) is connected by ducts to the venom glands under the thigh muscles (below).

venom gland

venom duct

spur

hunted for its pelt, animals were stunned by a heavy caliber shot fired under them in the water and gun dogs were sent out to retrieve them. There are numerous stories of dogs being killed by the platypus recovering and spurring them in the muzzle.

The glands associated with the spurs increase and recede in size with the testes during the breeding season, so they are thought to be associated with breeding. An early naturalist suggested that the male used its spurs to subdue the female during mating. This has not been observed and it is now thought that the crural system is probably used by males in establishing territories or access to mates during the breeding season. Certainly, males are more aggressive towards each other (and towards any other captor) at this time of the year. They also may avoid each other by being active at different times of the day or by not having their home ranges overlapping during the breeding season. Social organization in the species is poorly understood. The spurs certainly represent a deterrent to predators, but their loss in females suggests that this is not their primary function.

---

for breathing, make the monotremes undoubtedly part of the class Mammalia.

THE PLATYPUS

This small amphibious animal, with its pliable duck-like bill, thick fur, and strongly webbed forefeet, is perhaps one of the most unusual animals alive today. The species is common over its present distribution in the streams, rivers, and lakes of much of eastern Australia, where individuals can often be seen diving for the small insect larvae and other invertebrates that make up their food supply. Between dives, animals spend a few minutes chewing the food collected in their cheek pouches, using the horny pads that replace the teeth in adults. Dives usually last for less than two minutes, and during them the groove that encloses both the ears and eyes is closed and the platypus locates its food using its bill. This organ has both sensitive touch and electrosensors, enabling the animal to locate its prey and to find its way around underwater at night and in turbid waters. Although individuals are most often observed around dawn and dusk, they can be active throughout the night and may, especially in winter, extend their feeding into daylight hours.

Once thought to be primitive because of a presumed poor ability to regulate its body temperature, the platypus has been found to be a competent homeotherm, maintaining a body temperature of 32°C (90°F) even when swimming

for extended periods in near-freezing water. The animal raises its metabolism to generate more heat in cold conditions. This process is made more efficient by good body insulation, including a coat of fine dense fur, which retains an insulating layer of air when the animal is in water. (This fur was once prized by humans and the species was widely hunted until its protection early this century.)

When it is out of water the platypus normally occupies a burrow, dug into the bank of a river. Here it is buffered from the extremes of outside temperatures, which can vary from around −15°C (5°F) in the winter to as high as 40°C (104°F) in the summer. It has been suggested that platypuses may hibernate. Researchers know that some individuals stay in their burrows for several days at a time during winter but radio-telemetry studies have so far failed to detect hibernation.

Not all females in a platypus population breed each year and neither males nor females appear to breed until at least their second breeding season. Females lay from one to three eggs, with two being the most common number. The eggs are thought to be incubated between the tail and body of the female as she lies curled up in a special nesting burrow of up to 30 meters (33 yards) in length. When the young hatch, they remain in this burrow for three to four months and are fed on milk by the mother throughout that time. The milk oozes out onto the fur from two nipple-like patches on the underside of the mother's body,

▲ *The small size of the platypus can be gauged from this photograph. This is a female, which grows to 55 centimeters (22 inches); the male can reach 63 centimeters (25 inches). When it is feeding at the surface, a platypus can submerge almost totally, with only the nostrils and the top of the head and back above the water.*

and it is presumed that the young take the milk up from the fur. The breeding season is extended, with mating from July to October. After a gestation period (thought to be around three weeks), incubation of around ten days, and the long period of suckling in the burrow, the young enter the outside environment between January and March each year. It appears that the breeding season may be earlier in the northern part of the species' distribution and later in the south, particularly in Tasmania. Tasmanian platypuses are also normally larger than those on the Australian mainland.

In captivity platypuses are known to survive for up to 20 years, but the longest recorded longevity in the wild is 15 years. Practically nothing is known of mortality in wild populations, although there is some predation by foxes and birds of prey. Juvenile dispersal occurs each year and it is assumed that this results in considerable mortality, especially in drier years when the amount of suitable habitat is reduced. The platypus carries a range of external and internal parasites, including its own unique species of tick, but it is not known if any of these contribute to mortality. The species has been bred only once in a zoo and, because individuals are easily stressed, it is difficult to keep them successfully under captive conditions.

ECHIDNAS

Long-beaked and short-beaked echidnas are animals with a snout modified to form an elongated beak-like organ. They have a long protrusible tongue and no teeth. In addition to normal hair, they have a number of special hairs on the sides and back which are modified to form sharp spines. The long-beaked species, at 60 to 100 centimeters (24 to 39 inches) in length and 6 to 10 kilograms (13 to 22 pounds) in weight, is larger than the short-beaked species, which is only 30 to 45 centimeters (11 to 18 inches) long and 2 to 7 kilograms (4 to 15 pounds) in weight. In both species only the male retains the spur on the ankle of each rear leg.

Although the area of distribution of the long-beaked echidna is poorly studied, the species is now considered endangered. The short-beaked echidna is distributed throughout mainland Australia and Tasmania, where its status can be regarded as common. In New Guinea it is still considered to be common, although both species are known to be preyed upon by humans for food.

Unlike the platypus, the ears and eyes of echidnas are not housed in the same groove; the ear opening (with little visible external ear) is well behind the eye. The snout and protrusible tongues are both used in feeding. The short-beaked echidna eats mainly termites and ants although insect larvae, beetles, and earthworms are also taken. It procures ants and termites by excavating the mounds, galleries, and nests of these insects with the large claws on its front feet. The echidna then picks up the ants or termites with its sticky tongue. It can push its elongated snout into small spaces and extend its tongue into small cavities to gain access to these insects. The generic term *Tachyglossus* actually means "swift tongue". The long-beaked echidna is chiefly a worm eater. It uses spines housed in a groove in its tongue to draw the worms into its mouth. In both species, mucous secretions make the tongue sticky and, in the absence of teeth, food material is ground between spines at the base of the tongue and at the back of the palate.

Little is known of the activities of the New Guinea echidnas, but in Australia echidnas can be active at any time of the day, although they seem to be less active and stay buried in soil or shelter under rocks or vegetation in extremes of heat or cold. Like the platypus, they are unable to tolerate high temperatures and will die of heat stress if shade is not available. The burrowing ability of the short-beaked echidna is legendary, with individuals able to burrow vertically down into the earth to disappear in less than a minute.

Echidnas are endothermic and, like platypuses, can regulate their body temperatures well above that of the environmental temperatures by raising their metabolism and using insulation—fur and fat in the case of the echidnas. In all three species of monotreme the temperature maintained is lower than that found in many other mammals, but is usually maintained within a few degrees of 32°C (90°F) while the animals are active. It is now known that the short-beaked echidna sometimes hibernates for two to three weeks during winter in the Australian Alps, when body temperatures of individuals can fall to 4 to 9°C (39 to 48°F).

▼ The long-beaked echidna is hairier and less spiny than the short-beaked, perhaps because it lives in the cooler highland areas of New Guinea. It is a worm-eater and probes for earthworms in the ground with its long beak. It then uses its long grooved tongue to grasp the worms and pull them into its mouth. The nostrils and mouth are at the end of the beak.

D. Parer & E. Parer-Cook/Auscape International

D. Parer & E. Parer-Cook/Auscape International

Little is known of the breeding cycle of the long-beaked echidna. In the short-beaked species, a pouch develops during the breeding season, into which one egg is laid. After about 10 days of incubation, the young hatches and is nourished on milk suckled from the milk patches in the pouch, the prodding of the young stimulating the milk to flow. Lactation lasts for up to 7 months, but once the young begins to grow spines (around nine weeks after hatching), it is left in a burrow to which the mother returns to feed it. As in the platypus, the breeding season is extended. Mating occurs between May and September, when courtship behavior may involve several males (2 to 7 have been observed) following a single female for up to 14 days. After digging and pushing contests between males around the female, one male remains to copulate. The length of the gestation period is not known exactly but is about 3 weeks. Like platypus females, not all adult females in a short-beaked echidna population breed each year but the reasons for this are unknown.

Both species of echidna are long-lived. One short-beaked echidna in the Philadelphia Zoo lived for 49 years, and a marked individual in the wild was found to be at least 16 years of age. An individual long-beaked echidna survived for 31 to 36 years in Berlin Zoo, through both of the world

wars, but nothing is known of the longevity of this species in the wild. Dingoes are known to prey on echidnas, in spite of the echidnas' ability to burrow and their armory of spines. Foxes, feral cats, and goannas take young from burrows during the suckling period, but perhaps the greatest mortality factor is the automobile. The role of parasites or diseases in mortality is largely unknown. Echidnas are readily maintained in captivity but rarely breed successfully under captive conditions.

TOM GRANT

D. Parer & E. Parer-Cook/Auscape International

▲ Echidnas dig straight down into the ground as a defense mechanism, until only the very tips of the spines are visible. This gives them some protection from predators, but dingoes can dig them out and eat them, spines and all. This photograph of a mother and her offspring is unusual as the young are normally kept in the burrow, to which the mother returns every few days.

◄ The short-beaked echidna's long sticky tongue can be pushed out of the mouth and into holes in termite mounds to lick up the occupants. Echidnas can be seen foraging on anthills, covered in ants. They seem to favor female ants, perhaps because females have more fat on them and therefore make tastier morsels.

MARSUPIALS

Although marsupials (Metatheria) and non-marsupial mammals (Eutheria) have followed separate evolutionary paths for at least 100 million years, there are few fundamental differences between them. They are grouped together as members of the Theria (in contrast to the Prototheria, of which the monotremes are the only living survivors). Although anatomists can point to certain differences between all marsupials and all eutherians, the two groups are distinguished mainly by their reproduction: marsupial young are born in a much more undeveloped state than eutherian young.

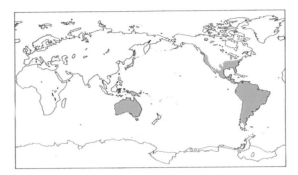

WHAT MAKES MARSUPIALS DIFFERENT
Other differences are demonstrated by many, but not all, marsupials. Marsupials have a greater maximum number of incisors (five pairs in the upper jaw, four in the lower) than eutherians (never more than three pairs in each jaw). In most marsupials, the first toe of the hindfoot is opposable to the other four, as in primates, and it always lacks a claw; many terrestrial marsupials, however, have lost this toe. In general, marsupials have a somewhat smaller brain than eutherians of equivalent size. Their body temperature and rate of metabolism are slightly lower than those of eutherians; it is tempting to regard marsupials as "inferior" in this respect but, if so, eutherians would have to be regarded as inferior to birds.

► *The woolly opossum (top) spends its life in the forest canopy, feeding mainly on fruits. The large eyes face forward, giving it a monkey-like appearance. Although most marsupials can swim, the yapok or water-opossum (bottom) is the only truly amphibious species, and it has webbed hindfeet. It feeds on invertebrates, fishes, and amphibians on the bottom of streams or lakes. Its eyes are shut when it is underwater and it locates its prey by probing with its sensitive fingers.*

CLASSIFYING MARSUPIALS

Classification of the marsupials is still a matter of debate but current expert opinion is that the modern fauna comprises two distinct assemblages: the Ameridelphia, restricted to the Americas, and the Australidelphia, comprising all the species found in Australia, New Guinea, and nearby islands plus one species from South America.

There is also growing consensus on the recent subdivisions of families of marsupials into the 19 families recognized here. There is not yet consensus, however, on whether to treat all marsupials as one order or to divide them into several orders, but as the evidence suggests that all marsupials are more closely related to each other than to any other mammals, the conservative course of recognizing one order until the weight of evidence points to a single alternative is to be preferred.

AMERICAN MARSUPIALS

There are approximately 75 species of living Ameridelphians, all of which are referred to as opossums, although they comprise two distinct families: Didelphidae (American opossums) and Caenolestidae (shrew-opossums).

American opossums These "true" opossums range from the cat-sized Virginian opossum *Didelphis virginiana* to the mouse-opossums, genus *Marmosa*. Most are omnivorous and able to climb,

but many spend more time on the ground than in trees, and the tails of the more terrestrial species are shorter and less prehensile. The carnivorous short-tailed opossums, genus *Monodelphis*, are mainly terrestrial, while the woolly opossums, genus *Caluromys*, are strongly arboreal and eat mostly fruits and nectar. The lutrine opossum *Lutreolina crassicaudata* is a predator and enters water to swim after prey. The yapok *Chironectes minimus* swims below the surface to take prey on the bottom.

Shrew-opossums The seven species of shrew-opossums are survivors of a group well known from fossil remains of the past 35 million years. Now restricted to cool mist-forests of the Andes, these mouse-sized to rat-sized marsupials are characterized by a pair of long, chisel-edged incisors that project forward from the lower jaw and are used to stab the large insects and small vertebrates upon which they prey.

Colocolo The colocolo or monito del monte *Dromiciops australis*, from the cool rainforests of southern Chile, is the only surviving member of the Microbiotheriidae family which has a long fossil history. The size of a small rat, it feeds mainly upon insect larvae and pupae. Most experts believe the colocolo to be related to the American opossums but increasing evidence suggests an affinity with the Australidelphia: if so, the colocolo

▲ In many respects, the Tasmanian devil is the marsupial equivalent of the hyena: it takes some live prey but is essentially a scavenger. Its powerful jaws enable it to completely consume a dead sheep, including the skull.

▲ *The red-tailed phascogale is one of the few dasyurid marsupials to spend much time in trees hunting prey.*

would constitute an important evolutionary link between the marsupials of the two continents.

AUSTRALIAN MARSUPIALS

The 200 or so living species of Australidelphians are much more diverse than the living Ameridelphians. The 16 families fall into 4 very distinct assemblages: the carnivorous marsupials; the marsupial mole; the bandicoot group; and a large group known as diprotodonts, which includes the koala, wombats, possums, and kangaroos.

Carnivorous marsupials

These comprise three families: the Dasyuridae (dasyurids), Thylacinidae (thylacine), and Myrmecobiidae (numbat). They have three or four pairs of narrow, pointed upper incisors and three pairs of similar shape in the lower jaw: the hindfeet have four or five toes, none of which are joined.

Dasyurids Dasyurids range from the tiny, flat-headed planigales, genus *Planigale*, weighing about 4 grams (less than ¼ ounce) and with a head and body length less than 6 centimeters (2½ inches),

to the terrier-sized Tasmanian devil *Sarcophilus harrisii*. Most are terrestrial but, although the tail is never prehensile, many can climb, and several, particularly the phascogales, genus *Phascogale,* are actively arboreal. There is very little variation in body shape except in the kultarr *Antechinomys laniger,* which has very long hindlegs and tail, which it employs in an unusual bounding gait.

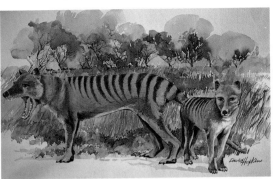

▲ *Because it sometimes attacked sheep, the thylacine was persecuted in Tasmania and the last known individual died in a zoo in 1933.*

▶ *The spotted-tailed quoll (top) is a cat-sized predator. Although an agile climber, it spends more time on the forest floor than in trees. The bilby (bottom left) lives in the Central Australian deserts. The silky fur provides thermal insulation, while the long ears probably act as radiators of excess body heat. The numbat (bottom right) is the only marsupial that feeds mainly on termites.*

Thylacine The thylacine or "Tasmanian tiger" *Thylacinus cynocephalus* probably became extinct in the 1940s. With a head–body length of more than 1 meter (3 feet), it was the largest carnivorous marsupial to have survived into historical times. Except for its broad-based and rather inflexible tail, it has the general conformation of a wolf, although it was striped rather like a tiger.

Numbat Sole member of the family Myrmecobiidae, the numbat *Myrmecobius fasciatus* is the only marsupial to be habitually active during the day. With the aid of a long sticky tongue, it feeds exclusively on ground-dwelling termites; its teeth are reduced to simple pegs.

Marsupial mole

The marsupial mole *Notoryctes typhlops* spends its life moving below the desert sands, feeding on burrowing insects and reptiles. It is the only living member of the family Notoryctidae. Very little is known of its behavior, and its relationship to other marsupials remains a mystery.

The bandicoot group

A dozen or so omnivorous species from Australia and New Guinea have an arrangement of teeth rather similar to that of the carnivorous marsupials, but they differ from these in many other ways, notably in having the second and third toes of the hindfoot fused together to form what looks like one toe with two claws. The body is compact, the snout elongate and pointed. Bandicoots are omnivorous, digging for insects, roots, and fungi with their clawed forefeet. The group is divided into two families: the mainly Australian Peramelidae and the mainly New Guinean Peroryctidae.

M.W. Gillam/Auscape International

Peramelids These comprise the "typical" bandicoots of the genera *Perameles* and *Isoodon*, which have bristly hair, short ears and tail, and a high rump. The legs are powerful, rather short, and employed in a bounding gait. The bilby, or rabbit-eared bandicoot, has long silken fur, long ears and a long furred tail. It spends the day in a deep burrow. The pig-footed bandicoot, extinct since the nineteenth century, had relatively long legs and probably ran on the tips of its hoof-like toes.

Spiny bandicoots These look like the "typical" bandicoots but have shinier and more spiny hair. They differ considerably from peramelids in their dentition and in details of cranial anatomy.

▲ Remarkably similar to the golden moles of Africa, the Australian marsupial mole is a striking example of convergent evolution. It does not make a tunnel but "swims" through the sand, which collapses behind it.

◄ The northern brown bandicoot uses its strongly clawed forelimbs to dig for succulent roots and underground insects. The powerful hindquarters and long hindfeet are employed in a bounding gait — a sort of "bunny hop".

Robert W.G. Jenkins/Australasian Nature Transparencies

▶ *"Koala" is said to be an Australian word meaning "doesn't drink", but it is more likely to have meant "biter". Koalas are adapted to eat the leaves of eucalypts, which contain large quantities of poisonous proteins. Despite their apparent sluggishness in daytime, at night they are active and agile climbers.*

▼ *With a head and body length of no more than 8 centimeters (3 inches), the feathertail glider is the smallest of the volplaning mammals. The featherlike tail is used to steer it in flights of more than 20 meters (65 feet).*

J. Cancalosi/Auscape International

Jean-Paul Ferrero/Auscape International

Diprotodonts

More than half of the Australidelphian marsupials belong to this diverse group, characterized by having only one pair of well-developed incisors in the lower jaw, and by the fusion of the second and third toes of the hindfoot (as in the bandicoot group) to form a two-clawed grooming comb. Most are herbivorous, but many eat insects and some lap plant exudates.

Wombats The three living species of wombats in the family Vombatidae are stocky, grazing animals with very short tails and a somewhat bear-like appearance. They walk on the soles of their feet, which have short toes armed with strong claws. Wombats excavate long burrows in which they sleep during the day.

Koala The koala *Phascolarctos cinereus* is the only living member of the family Phascolarctidae. A long-limbed, almost tailless arboreal marsupial, it feeds on the leaves of certain eucalypts. It is related, but not closely, to the wombats.

Cuscuses and brushtail possums The family Phalangeridae includes the cuscuses, genus *Phalanger,* and others; the cuscus-like scaly-tailed possum *Wyulda squamicaudata;* and the brushtail possums, genus *Trichosurus.* Except for the brushtail possums, they have a strongly prehensile tail, much of the free end of which is bare. They are arboreal, feeding mainly upon leaves, supplemented by buds, fruits, and shoots.

Ringtail possums and greater glider All but one of the members of the family Pseudocheiridae are ringtail possums, so called in reference to the long, slender, short-furred, and strongly prehensile tail: the rock ringtail *Pseudocheirus dahli,* which makes its nest on the ground, has a shorter tail than the other, exclusively arboreal, species.

The greater glider *Petauroides volans,* largest member of the family, is exceptional in having a membrane between the elbow and ankle on each side of the body, which is employed in gliding from tree to tree: the long, well-furred tail is only weakly prehensile. All members of the family feed mainly on leaves, the greater glider being restricted to those of eucalypts.

Wrist-webbed gliders and striped possums The family Petauridae includes four gliding species in the genus *Petaurus:* they differ from the greater glider in having a membrane that extends from the wrists (not the elbows) to the ankles. Leadbeater's possum *Gymnobelideus leadbeateri* resembles the sugar glider *Petaurus breviceps,* but lacks a gliding membrane. All feed on plant exudates and insects. Four species of striped possums, genus *Dactylopsila,* are characterized by a skunk-like coloration of black and white, and by having a very long, slender fourth finger; they feed exclusively on wood-boring insects.

Pygmy-possums The five species in the family Burramyidae include four arboreal pygmy-possums, genus *Cercartetus,* weighing 7 to 30

▼ The greater glider (below) is, at 1 meter (3 feet), the largest of the volplaning marsupials. Like the koala, the greater glider can digest the toxic leaves of gum trees.

grams (¼ to 1 ounce). They have long, slender, prehensile tails which are used to assist their agile climbing in shrubs and trees in search of nectar, pollen, and insects. The fifth species, the mountain pygmy-possum *Burramys parvus*, which at 40 grams (1½ ounces) is the largest member of the family, is predominantly terrestrial and feeds on insects, green plants, and seeds. It is the only Australian marsupial to live above the snowline.

Feathertails The two species in the family Acrobatidae are characterized by a row of long, stiff hairs projecting horizontally on each side of the tail. In the tiny feathertail glider *Acrobates pymaeus*, weighing about 12 grams (½ ounce), these hairs are so closely apposed that the tail resembles the flight feathers of a bird; the tail is employed, in combination with a membrane between the elbows and the knees, in gliding from tree to tree. In the larger feathertail possum *Distoechurus pennatus*, which lacks a gliding membrane, the lateral hairs are not closely apposed.

Honey possum The honey possum *Tarsipes rostratus* is the only member of the family Tarsipedidae. It is a small marsupial: males weigh only 9 grams (⅓ ounce), and females, 12 grams (½ ounce). The honey possum uses its long, fringed tongue to feed exclusively upon nectar.

▲ The squirrel glider (top left) is one of several gliding marsupials in which the gliding membrane extends from the ankle to the wrist. It inhabits the canopy of open forest. The striped possum (top right) of New Guinea and the tropical rainforests of Australia digs into the tunnels of wood-boring insects with its chisel-like lower incisors and extracts the insects with its long, slender fourth finger. Most cuscuses (above, bottom) are found in New Guinea and nearby islands. With strongly grasping hands and feet and a powerfully prehensile tail, they move deliberately through the rainforest canopy. The face tends to be flat and the eyes face forward; early European explorers often mistook them for monkeys.

Rat-kangaroos Members of the family Potoroidae retain some relatively unspecialized characters that have been lost in the closely related "true" kangaroos of the family Macropodidae. Most "primitive" of the group is the musky rat-kangaroo *Hypsiprymnodon moschatus*, which has a head and body length of only 23 centimeters (9 inches). It is notable for having five toes on the hindfoot (the other potoroids and "true" kangaroos have four), and for its bounding (rather than hopping) gait when it moves fast.

Other members of the group are the potoroos, genus *Potorous,* and bettongs, genus *Bettongia,* and others, most of which feed on underground tubers, bulbs, corms, and fungi, sometimes supplemented by green plant material. The tail, employed as a balance when hopping, is weakly prehensile and is used to carry nesting material.

Kangaroos and wallabies Probably the most recently evolved group of the marsupials, the Macropodidae comprises 11 genera and more than 50 species of wallabies, kangaroos, and tree-kangaroos. All, except the tree-kangaroos, genus *Dendrolagus,* share a similar body shape, with short forelimbs and large, powerful hindlimbs, long hindfeet, and a very large fourth toe. When moving fast, they hop on the hindlimbs, using the long and rather inflexible tail to provide balance. Rock-wallabies, genus *Petrogale,* which have somewhat shorter feet than the typical terrestrial kangaroos, are very agile denizens of cliffs and rock piles.

Tree-kangaroos, which evolved from terrestrial macropodids, have very long, strong forelimbs and shortened, broad hindfeet: they can walk along a horizontal branch or climb vertically by gripping a branch or stout vine with the claws of all four feet. All members of the family are herbivorous: in general, the more "primitive" species tend to be browsers, the more "advanced" to be grazers.

HOPPING, GLIDING, AND SWIMMING

The familiar hopping gait of kangaroos probably evolved from the bounding gait that has been retained by the diminutive and primitive musky

▲ *One of the largest macropods, the western grey kangaroo feeds on grasses. Long after it is able to move about on its own, the young of the western grey kangaroo enters its mother's pouch to avoid danger, to sleep, or to travel.*

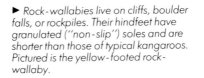

▶ *Rock-wallabies live on cliffs, boulder falls, or rockpiles. Their hindfeet have granulated ("non-slip") soles and are shorter than those of typical kangaroos. Pictured is the yellow-footed rock-wallaby.*

▶ Goodfellow's tree kangaroo (top right) is the brightest-colored of the tree kangaroos, which have evolved from terrestrial kangaroos. In adaptation to climbing, their forelegs became longer and stronger, while the hindfeet became shorter and broader. In many respects the musky rat-kangaroo (centre) provides evidence of the evolutionary origins of kangaroos from possums. It retains an opposable first toe; it bounds rather than hops; its tail is prehensile; and it gives birth to twins. The red kangaroo (bottom left) is the largest of the living marsupials. Old males may be 2 meters (8 feet) high when sitting, considerably more when propped on the tail. Females, and about 30 percent of males, are not red but bluish gray. The large-eyed forest wallaby of New Guinea (bottom right) inhabits the dimly lit rainforest floor. The tail, which has a scaly patch at the tip, is bent into a rightangled prop when the wallaby sits.

Jean-Paul Ferrero/Auscape International

▲ When hopping fast, a kangaroo uses less energy than a four-legged animal of the same size, moving at the same speed. A hopping kangaroo operates like a pogo stick.

▶ Most male kangaroos and wallabies attempt to dominate several breeding females. This leads to competition between males, often leading to fierce combat.

Hans & Judy Beste/Auscape International

rat-kangaroo. The hindlimbs became increasingly larger than the forelimbs and the hindfoot became longer, providing a very effective means for fast, hopping locomotion. At equivalent speed, and making allowance for the differences in weight, a hopping kangaroo uses less energy than a running dog or horse.

But such specialization is not without cost — a typical kangaroo is unable to walk. When moving slowly, it raises the hindquarters on a tripod formed by the forelimbs and the tail pressed down to the ground, then swings both hindlimbs forward and always together. Terrestrial kangaroos cannot move each hindleg independently while supporting the body, although they can and do kick them alternately when swimming. Tree-kangaroos, which have undergone a secondary shortening of the hindfeet, can also move their hindlegs alternately when walking along a branch.

In typical kangaroos, the tail is not very flexible but is moved up and down to assist with balance during the hopping gait and acts as a fifth limb to support the body when moving slowly. In the more primitive rat-kangaroos the tail is moderately prehensile in the vertical plane, and is used to carry bundles of nesting material.

The ability to glide has arisen independently in three families of marsupials: the Pseudocheiridae, Petauridae, and Acrobatidae. In each case, this involves a membrane of skin between the forelimbs and hindlimbs, which, when the legs are extended, expands into a rectangular, kite-like airfoil. Leaping from a high tree, a glider can volplane quite long distances, steering itself by altering the tension of the membrane on either side, balancing with the outstretched tail, and finally orienting the body vertically to land on the trunk of another tree with all four feet. No other gliding mammal has anything comparable to the tail of the feathertail glider: each side bears a thin row of stiff, closely packed hairs, all of the same length, forming a structure very similar to the vane of a feather.

Although it seems that most marsupials are able to swim when necessary, only the yapok can be regarded as truly aquatic. With alternate strokes of its webbed hindfeet, this Central American species swims to the bottom of a pond or stream with its eyes closed, feeling with its long, spatulate fingers for living prey, which it grasps in its mouth and takes to the shore to be eaten. The rear-opening pouch of a female yapok is closed by a strong sphincter muscle and sealed with water-repellent secretions when the animal is swimming.

The most extreme locomotory specialization of any marsupial is seen in the small, sausage-shaped marsupial mole. It is blind, lacks external ears, and has a horny shield over the snout and surrounding the nostrils. The limbs are short, with very strong bones and powerful musculature, and the spade-like forefeet have two immense, triangular claws on the third and fourth digits, with smaller claws on the other digits. The forefoot is employed in digging away soil in front of the animal and pulling it forward. The hindfoot, which has four short claws and, uniquely among marsupials, a claw-like structure on the first toe, is used to kick sand backward from the body. The marsupial mole does not construct a tunnel but "swims" through sand, usually at a depth of 10 to 20 centimeters (4 to 8 inches), but sometimes descending to a depth of 2 meters (6½ feet) or more.

The feathertail glider, the honey possum and pygmy-possums, genus *Cercartetus*, are very small marsupials that climb by gripping with the expanded tips of their fingers and toes. Except on the conjoined second and third toes of the hindfoot, the claws are reduced and nail-like, lying above the tips of the digits. The pads on the fingers and toes of the feathertail glider are microscopically grooved, like those of geckos, enabling them to cling to a smooth surface such as a vertical sheet of glass and even to hold themselves, albeit briefly, to the underside of a horizontal sheet.

SPECIALIZED DIETS

The first marsupials probably fed mainly on insects and other small invertebrates, as do the least specialized of the living opossums, dasyurids, and possums. It was a comparatively simple step in evolution for the larger dasyurids and the thylacine to graduate to preying upon vertebrates. An extreme adaptation to insectivory is the long sticky tongue of the numbat. Another is seen in striped possums, which expose wood-boring insect larvae with their protuberant lower incisors, then extract these with the claw of the very long, thin fourth finger. The Madagascar aye-aye—a primate—feeds in a similar manner.

No marsupial feeds entirely upon fruits, but many opossums, possums, and even some dasyurids, include these in an omnivorous diet. Gliders of the genus *Petaurus* feed partly on insects but also on the gum and sap of trees, promoting the flow of sap from eucalypts by cutting grooves into the trunk with their sharp lower incisors (a method of feeding also used by the pygmy marmoset). Bandicoots and the bilby dig in the earth for insect larvae, succulent tubers, bulbs, and corms (subterranean stems). Rat-kangaroos also make shallow excavations in search of underground fungi, but some also eat insects and the soft parts of green plants, particularly shoots.

Many diprotodont marsupials eat leaves. Tree-

Jean-Paul Ferrero/Auscape International

▲ The tiny honey possum feeds exclusively on nectar and has a long, brush-tipped tongue.

▼ The numbat uses its long, sticky tongue to lick up termites.

Dick Whitford/Australasian Nature Transparencies

kangaroos and cuscuses feed mainly on the soft, broad leaves of rainforest trees, supplemented by fruits, although cuscuses also eat large insects, eggs, and nestling birds. Brushtail and ringtail possums rely to a large extent upon the tough leaves of eucalypts, while the koala and greater glider rely exclusively on this source, which, although abundant, is not very nutritious and has a high content of toxic substances. Many wallabies browse on the leaves of shrubs and low trees but may also eat some grasses. The only marsupials to feed mainly on grasses are wombats, some species of wallaby, and *Macropus* kangaroos.

Because mammals lack the enzymes necessary to break down plant fiber into its constituent sugars, leaves and grasses are difficult to digest, although this can be managed with the aid of microorganisms if these are given sufficient space and time to act. Most marsupials promote the digestion of plant fiber by diverting chewed food into the cecum (a capacious diverticulum of the intestine) where it remains for some time, undergoing microbial fermentation before being returned to the intestine. Others, notably the wombats, chew grass very finely and pass it very slowly through an elongated large intestine. The koala is notable for having, proportionately, the largest cecum of any mammal, as well as a very long large intestine. Ringtail possums have a large cecum, but it appears to be inadequate for complete digestion of plant material — the contents of the cecum (so-called "soft feces") are evacuated once a day and re-eaten; this material passes through the body once more, the unassimilated fraction being voided as hard fecal pellets.

Kangaroos manage the digestion of tough grasses in a quite different manner. Finely chewed food is retained for microbial digestion in a compartment of the stomach before it passes on to the rest of the alimentary canal. Although comparable in some respects with the ruminant digestion of sheep and cattle, the details of anatomy and function are quite different in kangaroos, since rumination (chewing the cud) is not an essential part of the process.

RESTING IN THE COLD

Under cold conditions, much of the energy that a very small mammal obtains from its food is expended in maintaining its body temperature. Since cold is greatest in winter, when food also tends to be scarce, such mammals face considerable difficulties. One solution to the problem is to hibernate, permitting the temperature of the body to drop almost to that of the surrounding environment and to enter a state of almost suspended animation. This reduces both the rate of heat loss and the need for food.

There is some debate about whether any marsupials truly hibernate, but the sugar glider, the honey possum, the feathertail glider, most pygmy-possums, and some small dasyurids are known to respond to cold or to a shortage of food by becoming torpid overnight or for periods of a week or so. A torpid animal "switches off" its thermal regulation and cools down to a temperature a little above that of its surroundings. Its heartbeat and metabolism slow down accordingly. Most species that become torpid also conserve body heat by huddling together in a communal nest.

DAYTIME SHELTERS

Most marsupials sleep during the day in a nest of some sort, usually of dry foliage. A few species,

▼ *This sugar glider is about to land on the trunk of a tree, which it will grip with its hands and feet. Note the thumb-like first toe, ready to grip against the other toes.*

C. & S Pollitt/Australasian Nature Transparencies

such as pygmy-possums and gliders of the genera *Petaurus* and *Acrobates,* construct compact woven nests, which are usually situated in a tree hollow but are sufficiently robust to be self-supporting if built in a forked branch. Most other marsupials make a less structured nest in a tree hole, hollow log, or other crevice; planigales and some other very small, flat-headed dasyurids make nests during the dry season in the deep cracks that develop in clay soils. Bandicoots typically make a scrape in the ground and heap vegetation over it to make a nest that appears to be without structure

but may have a definite entrance and exit. Most rat-kangaroos build nests, often against the trunk of a tree or under a bush; rock-wallabies and the wallaroo *Macropus robustus* escape the heat of the day by sheltering under rock overhangs or in caves, while other wallabies and kangaroos shelter in the shade of trees or bushes. Wombats, the bilby, the burrowing bettong, and most desert-dwelling dasyurids make nests in burrows. Tree-kangaroos, the larger cuscuses, and the koala have neither nests nor dens. Tree foliage provides shelter from sunlight, but they have no protection from rain.

▲ Dunnarts look rather like small rodents and were once called "marsupial mice." A better comparison would be with shrews, for dunnarts are fierce little predators. This one is eating a desert grasshopper.

Dave Watts/Australasian Nature Transparencies

▲ *Wombats produce one young, which is accommodated in a backward-opening pouch. Once able to move about, a young wombat follows its mother at heel.*

BIRTH AND ATTACHMENT

As in other mammals, the right and left ovaries of a female marsupial shed their unfertilized eggs into corresponding right and left oviducts (fallopian tubes). In other ways, however, the anatomy of the marsupial female reproductive system is quite different. Each oviduct continues into a separate uterus and thence into a lateral vagina. The two vaginas join a median tube (the urogenital sinus) that opens, together with the rectum, into a shallow cavity or cloaca. (Monotremes are usually defined as mammals that possess a cloaca, but this does not distinguish them from female marsupials, which also pass feces, urine, and newborn young out through a single cloacal opening.)

This reproductive system would appear to be a simpler, more primitive arrangement than is found in eutherian mammals, where the oviducts open into a single uterus (sometimes paired at its apex) and thence into a single vagina, which opens to the exterior. However, the marsupial condition is also more specialized — indeed unique — in having a third, median, vagina. This is a passage connecting the two uteri to the urogenital sinus and through which the young are born. In most marsupials it is a temporary structure that develops just before parturition and disappears after the birth has taken place, but in kangaroos and the honey possum it remains permanently once it has been formed for the first birth.

Male marsupials differ from male eutherians in having the scrotum in front of the penis. The penis is forked in many species, presumably to direct semen into each of the lateral vaginas. Sperms pass up the vaginas and into the oviducts, where they fertilize one or more eggs. As a fertilized egg passes down the oviduct, it becomes enclosed in a very thin shell-membrane, a vestige of the ancestral eggshell. This disintegrates in the course of gestation and the embryo becomes attached to the wall of the uterus. While enclosed in a shell-membrane, a marsupial embryo is nourished by fluids secreted into the cavity of the uterus from glands in its wall. Embryos of some species make only a tenuous connection with the uterine wall: these receive most of their nourishment from the uterine secretions. Those that develop a significant placenta are able to supplement this source with materials passing almost directly from the maternal blood into the blood of the embryo.

Gestation is short in marsupials, ranging from about 9 days in the eastern quoll *Dasyurus viverrinus* to 38 days in the eastern gray kangaroo *Macropus giganteus.* Newborn marsupials are very small. The female honey possum weighs approximately 12 grams (½ ounce) and gives birth to young weighing about 5 milligrams ($\frac{1}{5,000}$ ounce); the female red kangaroo *Macropus rufus,* weighing approximately 27 kilograms (60 pounds), produces a neonate weighing only 800 milligrams (¹⁄₃₀ ounce) (0.003 percent of the maternal weight, compared with 5 percent in humans).

All marsupial neonates are remarkably similar: the skin is bare, thin, and richly supplied with blood (possibly acting as a respiratory surface); the eyes and ears are embryonic and without function (although the ear may be sensitive to gravity); and the hindlimbs are short, five-lobed buds. However the nostrils are immense; the sense of smell is well developed; the mouth is large, with a large tongue; the lungs and alimentary canal are functional; and the forelimbs are disproportionately large,

Jean-Paul Ferrero/Auscape International

▲ *Firmly attached to one of its mother's four teats, this red kangaroo pouch embryo is more than a month old. Although still blind and hairless, it has hindlimbs and a tail.*

powerful, and equipped with needle-sharp claws.

Birth is rapid—a simple "popping out". Immediately it is free of its embryonic membranes, the neonate begins to move toward a teat, waving its head from side to side and dragging itself by alternate movements of each forelimb through the forest of hairs on its mother's belly. Once it has located a teat, the neonate takes it firmly into its mouth, the tip of the teat expanding to fill the mouth cavity. All of this is accomplished in no more than a few minutes.

A large glottis at the back of the neonate's mouth shuts off the teat from the air passage between the nostrils and the lungs, permitting it to breath while suckling, without danger of choking. Firmly attached to a teat, the neonate settles down to a period of passive growth and development, which is considerably longer than gestation. During this period it is called a "pouch embryo".

Birth and attachment to a teat have been most intensively studied in kangaroos, which have a deep, forward-opening pouch. The neonate must therefore climb upward to the lip of the pouch, then clamber (or tumble) down into it to locate a teat. But kangaroos are not typical marsupials since, apart from possums, they are the only marsupials to have a pouch that opens anteriorly. In other marsupials the pouch either opens to the rear or consists of little more than lateral folds around the mammary area; a pouch may even be lacking in some small marsupials. Obviously, a rear-opening pouch is more convenient for a newborn marsupial and this is probably the primitive arrangement.

Kangaroos are also misleading models because, although they have four teats, they normally give birth to only one young at a time. Among opossums, litters of six or more are common and the pale-bellied opossum *Marmosa robinsoni* sometimes carries a pouch young on each of its fourteen teats. From six to ten young are carried by some dasyurids; bandicoots commonly carry three or four; and pygmy-possums may have litters of from four to six.

Some marsupials give birth to more young than can be accommodated on their teats. A Virginian opossum is known to have produced an excess of at least nine, while the eastern quoll and the kowari can produce at least three extra. Little is known of the extent of such overproduction, since most supernumerary young are probably eaten or lost in the nest litter and they are unlikely to be noticed unless birth takes place on a smooth, clean surface.

In various eutherian mammals, including certain seals and bats, a fertilized egg develops into a tiny, hollow ball of cells and then becomes quiescent for a period. Known as embryonic diapause, this process also occurs in most kangaroos (although not in potoroids), some pygmy-possums and, possibly, the honey possum. In most kangaroos and wallabies the embryo from

a female's first mating passes through a normal gestation, is born, and attaches to a teat. Within a day or so of that birth, the female mates for a second time but the embryo from this mating becomes quiescent at the blastocyst stage and remains so until the first young is about to give up its attachment to its teat. The blastocyst then resumes normal development and, around the time of its birth, a third mating takes place and so the process is continued. Thereafter, throughout her reproductive career, the female will normally be carrying one blastocyst, one pouch embryo, and one young which is still suckling but able to move in and out of the pouch. The mammary gland of a teat that is supplying a pouch embryo produces milk of quite different composition from a teat that is supplying its older sibling.

Female marsupials play a passive role during the birth and teat-attachment of their young. Care of the pouch embryos is limited to cleaning the pouch or mammary area. Species that produce numerous offspring usually lack a sufficiently large pouch and, once their young have detached themselves from the teat, they are left in a nest during the night while the female is foraging. During the day she sleeps with them, suckles them, and grooms them. Young that stray a short distance from the nest may be retrieved in response to their distress calls but this behavior is not universal. When the young are fully formed and capable of independent movement, they may follow the mother while she is foraging or, more commonly, cling to her fur with their teeth and claws — creating a very considerable burden.

The larger marsupials tend to have capacious pouches and fewer (usually only one) young. In these species, the general pattern is for the young to inhabit the pouch long after they have become detached from the teat and, during the period of weaning, to use the pouch as a means of transport and a place to sleep; this is particularly the case in kangaroos, bandicoots, and wombats. In the case of the koala and the larger species of possum, partially weaned young are often carried on the mother's back, but they may be "parked" while the mother is on a long foraging excursion, and maternal care ceases at the end of weaning. Males are never involved in parental care.

Marsupials differ most notably from eutherians in the relatively small size of their newborn young —a marsupial mother's initial investment in reproduction is low. However, by the time that the young are weaned, the ratio between her weight and the weight of her offspring (whether a litter or a single individual) is similar to that of a eutherian mother. The difference is essentially a matter of how much time is spent developing in the womb. There is no evidence that marsupial reproduction is inferior to that of eutherians; as in other aspects, marsupials are simply different.

RONALD STRAHAN

Female eutherian
Genital aperture separate from anus; one vagina and one uterus.

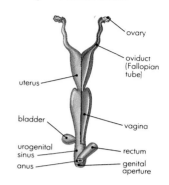

Non-pregnant female marsupial
Rectum and urogenital sinus open to outside through same (cloacal) aperture; two vaginas and two uteruses.

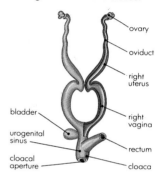

Pregnant marsupial
Median vagina develops, through which birth takes place.

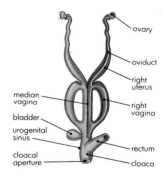

A.P. Smith/Australasian Nature Transparencies

▲ *Many male marsupials, like the sugar glider shown here, have a forked penis.*

ANTEATERS, SLOTHS & ARMADILLOS

ORDER XENARTHRA
- 4 families • 13 genera
- 29 species

SMALLEST & LARGEST

Lesser fairy armadillo
Chlamyphorus truncatus
Head–body length: 12.5–15 cm
(5–6 in)
Tail length: 2.5–3 cm (1–1⅕ in)
Weight: 80–100 g (2⅖–3½ oz)

Giant anteater *Myrmecophaga tridactyla*
Head–body length: 100–120 cm
(40–48 in)
Tail length: 70–90 cm (27–36 in)
Weight: 20–40 kg (44–88 lb)

CONSERVATION WATCH
!! The maned sloth *Bradypus torquatus*, lesser pichiciego *Chlamyphorus truncatus,* and giant armadillo *Priodontes maximus* are listed as endangered species.
! The *Chaetophractus nationi*; greater pichiciego *Chlamyphorus retusus*; *Dasypus pilosus*; Brazilian three-banded armadillo *Tolypeutes tricinctus*; and giant anteater *Myrmecophaga tridactyla* are listed as vulnerable species.

▼ *The collared anteater or tamandua climbs in trees with the aid of a prehensile tail. The three recurved claws on each forefoot are used to open ant and termite nests.*

The order Xenarthra comprises a bizarre group of animals that radiated in South America between the Paleocene and Pliocene epochs (65 to 2 million years ago). Four main lines evolved from this ancient stock: a lineage of armored terrestrial grazing herbivores; an arboreal (tree-dwelling) non-armored group specializing in browsing; a fossorial (burrowing) omnivore/insectivore group; and a lineage supremely adapted to feeding on ants and termites. The first lineage is represented today by the armadillos (20 species); the second, by the sloths (five species); the third is now extinct; and the fourth comprises the American anteaters (four species). Only a few of the weird and wonderful animals of this era are still to be found, but these relicts are some of the most fascinating and, ironically, some of the least-studied living mammals.

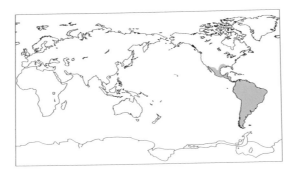

AN UNLIKELY GROUPING
It is difficult to imagine the connections between heavily armored, fast-moving armadillos which curl up like hedgehogs, and the lethargic, unkempt-looking sloths which spend most of their time hanging upside down in forest canopies, and the elegant movements of the anteaters which live on the plains as well as in trees. Yet, these species share a common origin, as well as many physical features.

Anteaters, sloths, and armadillos are strictly confined to the New World, having originated in North America. At the beginning of the Paleocene epoch, a land bridge connected the North and South American land masses, permitting the free passage of animals and plants between the two continents. In this way, South America was gradually colonized by the ancestors of many modern mammals.

When this umbilical land bridge was severed or submerged for at least 70 million years, those animals that had successfully colonized the southern "island" were able to evolve in complete isolation. This led to the evolution of many new orders, families, and genera with distinctive and often bizarre features.

ANTEATERS
Anteaters were probably one of the first groups of mammals to have reached the South American continent before it became an island, and would, as such, have evolved in the warm, moist seclusion of the tropical forest.

Giant anteaters
The giant anteater *Myrmecophaga tridactyla,* which ranges from 100 to 120 centimeters (40 to 48 inches) in length and 20 to 40 kilograms (44 to 88 pounds) in weight (though males are often 10 to 20 percent heavier than females), is gray with a black and white shoulder stripe. It is a familiar inhabitant of the South American savannas and open woodlands, but, on occasion, it may also venture into the tropical rainforest in search of its favorite foods—ants and termites. Endowed with only an average sense of vision, this magnificent animal possesses a highly attuned sense of smell; with its elongated, proboscis-like snout, it constantly sniffs the air and ground to locate potential prey. Scientific tests have shown that the

Haroldo Palo Jr./NHPA

Haroldo Palo Jr./NHPA

anteater's sense of smell is at least 40 times more acute than that of humans.

Upon locating a potential feeding place, the anteater digs an entrance into the nest with rapid movements of its toughened claws. The long, worm-shaped (vermiform) tongue is protrusible and is covered with sticky saliva secreted by enlarged glands situated at the base of the neck.

The giant anteater is the only member of the family Myrmecophagidae that is not predominantly arboreal. Although it can climb reasonably well, it is usually found on the ground moving with a characteristic rolling gait. Instead of supporting its body weight on the soles of its feet, as is the norm for most mammals, it is compelled to walk on the hard edges of each foot, turning the two largest claws (the second and the third) inward. The reason for this ungainly gait is that the lengthy, powerful claws are non-retractile and are, therefore, a hindrance to locomotion on the ground. However, the seemingly cumbersome movements of giant anteaters are deceptive, as they are able to outrun a human if necessary.

When the anteater is confronted by predators, its fearsome talons on the forelimbs serve another function apart from ripping open the concrete-like termite mounds—defense. When confronted with danger—for example, if surprised by a jaguar—the giant anteater will rear up to its full size on its hindlegs. The wielding, slashing actions of the

forelimbs seriously challenge further approaches from would-be predators, which generally retreat to avoid sustaining serious injury.

The astonishing evolutionary adaptations of the giant anteater and other members of the family Myrmecophagidae to such specialized diets are doubtless the result of millions of years of gradual development by ancestral forms in conditions of comparative isolation. Although mammals that feed on ants are to be found in most other zoogeographical regions, no other single species demonstrates the same degree of specialization as the giant anteater of South America. Others, particularly the African aardvark and pangolins, furnish an interesting example of convergent evolution for, like the anteaters of the South American tropics, they all make use of similar weapons to extract and trap their prey—strong, curving claws and a sticky, protractile tongue. They also possess the same type of muscular stomach, which is able to digest the toughened exoskeletons of insect prey.

Little is known about the social behavior of the giant anteater; we can say only that its gestation period is about 190 days.

Tamanduas and silky anteaters
The other representatives of the family Myrmecophagidae are the two species of collared anteater or tamandua—*Tamandua mexicana* in the

▲ *The immense claws and powerful forelegs of the giant anteater are used to dig into rock-hard termite mounds and to defend it against larger predators. It sleeps on its side, using the fan-like tail as a blanket.*

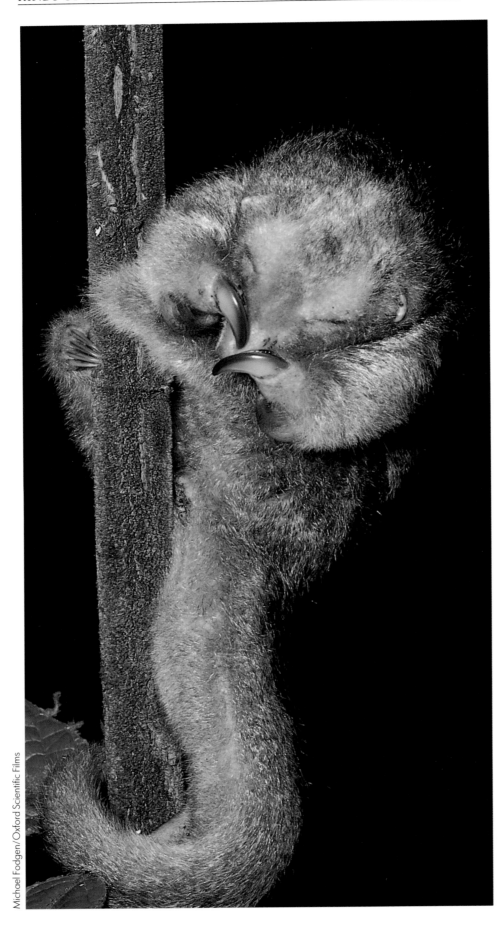

Michael Fodgen/Oxford Scientific Films

▲ The strictly arboreal silky anteater has only two functional toes on the forefeet, both armed with immense, hook-like claws. The furry tail is strongly prehensile.

north and *T. tetradactyla* in the south — as well as the two-toed or silky anteater *Cyclopes didactylus*. Unlike their much larger relative, these species prefer a forest or dense woodland habitat and, as might be expected in this habitat, are tree-dwelling animals. They seldom descend to the ground where, lacking the impressive defense of the giant anteater, they would be at the mercy of forest predators.

The two species of tamandua are about half the size of the giant anteater, with distributions that considerably overlap that of the giant anteater. Apart from color — the tamanduas' smooth fawn coats contrast strongly with the giant anteater's gray shaggy appearance — the main difference between these two genera is that tamanduas have a naked, prehensile tail which they use when climbing. Also, unlike the giant anteater, tamanduas are strictly nocturnal, using hollow trees as sanctuaries during daylight hours. But, as for the giant anteater, we know little about the tamanduas' reproduction, except that the gestation period is about 140 days.

The much smaller and even more secretive silky anteater is also nocturnal and strictly arboreal. Unlike its close relatives, the silky anteater rarely feeds on termites, preferring instead the delicate flavor of ants that inhabit the stems of lianas and tree branches. Its muzzle is proportionately shorter than that of its relatives, but its tongue is similar in structure and stickiness. Its fur is long, yellow in color, and silky in texture; the prehensile tail is longer than the body. Because of its superficial resemblance to tree-dwelling monkeys, this species is often referred to as the long-tailed tree monkey. Even though it may receive some degree of cover from its shady environment, the silky anteater is often threatened by aerial predators, such as the harpy eagle, hawk eagle, and spectacled owl. To defend itself against such attacks, it reacts like other members of its family by rearing on its hindlimbs and lashing out with its forelimbs.

SLOTHS

The sloths represent the survivors of an early mammalian radiation to South America that adapted to browsing. Strictly herbivorous, sloths are primarily leaf eaters and only rarely descend to the ground. In fact these animals are so highly modified for a specialized form of arboreal locomotion that they have almost lost the ability to move on the ground. The conversion of structural cellulose to simple sugars, as well as the detoxification of the many tannins found in leaves, was accomplished by the evolution of a chambered stomach where leaves are fermented with the aid of symbiotic protozoan and bacterial organisms.

The two families of sloths each contain a single genus, the three-toed sloth, family Bradypodidae, and the two-toed sloth, family Megalonychidae. Both are strictly New World species, confined to the tropical rainforests of Central and South America.

Sloths have rounded heads and flattened faces with tiny ears concealed under a dense, shaggy fur, and their hands and feet end in curved claws that are 8 to 10 centimeters (3 to 4 inches) in length. In both species the forelimbs are considerably longer than the hindlimbs, making movement on the ground awkward and ungainly. In fact, sloths' limbs are unable to support their body weight, and they move on the ground by slowly dragging the body forward with the forelimbs. They climb in an upright position by embracing a branch or by hanging upside down and moving along hand over hand. Sloths spend a considerable amount of time hanging upside down in the tree canopy — at times even sleeping in this position — where they resemble animated coat hangers.

The entire body is covered with a dense, soft fleece from which protrude larger, coarser tufts of hair. Each of these hairs is grooved, vertically and horizontally, and because of the almost constant humidity of the forest canopy, these narrow slits harbor a multitude of microscopic green algae. Consequently, although the basic color of the sloth's hair is gray or brown, the profusion of algae in the outer hairs gives the coat a distinctive greenish tinge, which effectively camouflages the animal from avian and terrestrial predators, such as the harpy eagle and jaguar.

Sloths are long-lived, solitary animals that appear to have no fixed breeding season. Following a gestation period of about one year the single young is born, high within the canopy, and initially helped to the teat by the mother, where it will feed for almost one month. The male does not participate in rearing the young. No fixed nesting

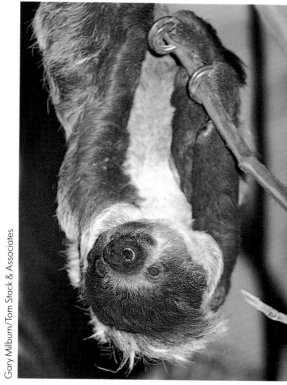

Gary Milburn/Tom Stack & Associates

◄ *The two-toed sloth lives in mountainous rainforests, where it feeds on the broad leaves of a small number of species of rainforest trees. While awake, it feeds almost continuously.*

site is used; instead the mother acts as a mobile nest with the infant clinging on to her. Following weaning the young remains with its mother, usually being carried, for a further six months. During this time the infant begins to feed from the same trees as its mother; adult sloths seem to have individually distinctive feeding preferences which they appear to pass onto their offspring.

◄ *The three-toed sloth seldom descends to the ground except to defecate—about once a week. It moves very clumsily on the forest floor because the muscles of its limbs are more adapted to hanging upside down than to supporting the body.*

Joe McDonald/Tom Stack & Associates

▶ *Sloths do not make nests or use dens. The single young must therefore be carried by its mother from the time of birth. The three-toed sloth, shown here, carries its young until it is about six months old.*

▼ *Sloths have long, shaggy hair. Each hair has a groove along its length and algae grow in this space. In wet conditions, the algae multiply and the hairs become green—perhaps contributing to camouflage.*

The vast majority of a sloth's active time is spent in feeding and, when full, an animal's stomach may account for almost one-third of its body weight, which ranges from 4 to 8 kilograms (9 to 18 pounds). In keeping with the animal's pace of life, digestion is a relatively slow process: up to a month may elapse before foodstuffs finally pass from the multichambered stomach to the small intestine.

Unlike most arboreal mammals, sloths descend from the trees to defecate on the ground, an exercise they engage in only once a week, but considering their lethargic way of life, this must consume a considerable amount of body energy. When on the ground three-toed sloths, at least, scoop out a small depression in the soil at the base of the tree before defecating. The site is apparently carefully selected, and, apart from serving as a useful olfactory signal to other sloths in the area, it has been suggested that sloths are actually fertilizing their favorite feeding trees by this action.

Such visits to the ground are also important for other species: the shaggy coat of most sloths harbors a wide range of insects, including moths, beetles, ticks, and mites. Although the exact interrelationship between these species and their "habitat" has been poorly studied, it is now known that at least one species of moth living in the fur lays its eggs on the dung that the sloth deposits at the base of the tree.

ARMADILLOS

When the first Spanish explorers set eyes on these small, scurrying creatures covered in hard, armor-like scales, they called these amazing animals "armadillios" (from the Spanish *armado*, meaning "armed creature"). As their name suggests, armadillos are armored with a flexible horny shield, backed by bone, over the shoulders, another over the hips, and a varying number of bands around the waist connecting the two. This body shell actually develops from the skin and is

Michael Fogden/Bruce Coleman Ltd

▲ Other armored mammals are protected by thickened skin or horny structures, but the armor of armadillos consists of plates of bone with a horny covering. The body of this Patagonian armadillo is mostly enclosed by eight hinged bands of these plates, the head and rear being covered by solid plates.

composed of strong bony plates overlaid by horn. Only the upper surface of the body and the limbs are fully armored; the ventral surface of the body is covered by a soft, hairy skin.

Less specialized feeders than the anteaters, the armadillos (family Dasypodidae) also eat their share of ants and termites from the savannas and woodlands of the South American tropics. Although most of the 20 species feed primarily on insects, they also eat a variety of invertebrates, small vertebrates, and vegetable matter. In terms of the number of species and the breadth of distribution — from Argentina, through Central America to Florida — this is the most successful edentate family.

Comparable in size to the giant anteater, the giant armadillo *Priodontes maximus* takes its food-finding task very seriously. Instead of fastidiously poking an elegant nose into a small corridor of the termite mound as, for example, an anteater would, the giant armadillo excavates a sizable tunnel directly into the mound until it reaches the heart of the colony, apparently oblivious to the bites of several thousand angry termite soldiers.

Most armadillos live in deep underground tunnels, which they excavate in sites not prone to flooding. Armadillos are generally nocturnal, often spending the day curled up with head and tail

folded neatly over the belly. Although at first appearance the toughened carapace of an armadillo would seem to provide adequate deterrent to most predators, it is, in fact, quite vulnerable to attack. The normal reaction for most species of armadillo is not to curl up immediately into a defensive position, but instead to flee, which is most effective in dense undergrowth, or to burrow, which is often accomplished at an astonishing rate. Two steppe-dwelling species, the fairy armadillo *Chlamyphorus truncatus* and the greater pichiciego *Chlamyphorus retusus,* block the entrance to their burrows by obstructing the passage with their armor-plated hindquarters. Only one species, the three-banded armadillo *Tolypeutes tricinctus,* does not burrow since it is able to roll into a tight ball.

As a rule, armadillos live in semi-arid and even desert zones; it is rare for them to venture into forests. The giant armadillo, however, does have a preference for forest habitat, although it also frequents the savanna in search of food. The most widespread species is the common long-nosed or nine-banded armadillo *Dasypus novemcinctus,* which is found in a range of habitats from North America to Argentina. The six species in this genus don't appear to be limited to a fixed breeding season; if environmental conditions are favorable, they will breed. The unparalleled success of

Francisco Erize/Bruce Coleman ltd.

members of this genus is probably due to their flexible reproductive behavior and diet, which allow them to exploit all but the most arid areas.

THREATS TO SURVIVAL

One distressing characteristic shared by the sloths, anteaters, and armadillos is that several species are severely threatened in their natural ecosystem. The magnificent giant anteater has already been exterminated from large areas of Brazil and Peru through direct hunting for trophies and also for the trade in live-animal collections. The skin of these animals is also put to some local use.

Probably the greatest single threat to the survival of these magnificent, bizarre creatures is, however, habitat destruction, which will undoubtedly have a severe impact on animals with such highly specialized, restricted feeding regimes. The vast savannas of Latin America are undergoing major habitat alterations, being used to feed the millions of cattle stocked by a number of ranches, while, elsewhere, the once extensive tracts of tropical rainforests are being severely eroded and irreparably damaged by deforestation for logging, slash and burn cultivation, and inundation as a result of hydroelectric dam construction. The

Francisco Erize/Bruce Coleman Ltd.

◄ *Many armadillos curl up when asleep or disturbed but, even in this posture, some parts of the body are exposed to predators. The three-banded armadillo curls into a completely enclosed ball.*

anteaters, sloths, and armadillos need not fear the challenge from more adaptable animal competitors in future years; their survival depends on the efforts of humans to protect the remaining vestiges of their habitats. Without such efforts, these highly specialized species are certainly doomed.

R. DAVID STONE

▼ *The armor of an armadillo has some chinks, which are found by defending soldier ants and termites when their nests are raided. This nine-banded armadillo is rubbing off ants that have penetrated its defenses.*

Jeff Foott/Bruce Coleman Ltd.

INSECTIVORES

ORDER INSECTIVORA
- 6 families • 65 genera
- *c.* 430 species

SMALLEST & LARGEST

Pygmy white-toothed shrew
Suncus etruscus
Head–tail length: 35–48 mm
(1³⁄₁₀–2 in)
Weight: 2 g (⁷⁄₁₀₀ oz)

Greater moonrat *Echinosorex gymnurus*
Head–body length: 26–45 cm
(10–18 in)
Tail length: 20–21 cm (7⅘–8³⁄₁₀ in)
Weight: 1,000–1,400 g (2–3 lb)

CONSERVATION WATCH
!!! 37 species are listed as critically endangered, including: Somali golden mole *Chlorotalpa tytonis*; dwarf gymnure *Hylomys parvus*; Malayan water shrew *Chimarrogale hantu*; and 17 shrew species in the genus *Crocidura*.
!! 48 species are endangered.
! 67 species are listed as vulnerable.

▼ *Solenodons, small animals about 300 millimeters (12 inches) long with a tail nearly as long again, investigate their surroundings with a long, mobile snout armed with sensory hairs. The forelimbs bear strong claws for digging and to hold prey, which rapidly succumbs to a toxic saliva injected by the solenodon's bite.*

The order Insectivora (insect eaters) is ancient, and as a group the "true" insectivores are generally considered to be the most primitive of living placental mammals and therefore representative of the ancestral stock from which modern mammals are derived. They are a diverse and ragged assemblage of animal groups that share a tendency—if not an extreme specialization—toward eating insects. Insectivores are probably best described as being small, highly mobile animals with long, narrow, and often elaborate snouts.

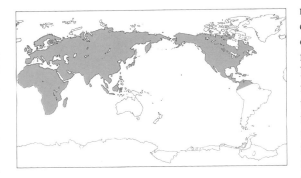

SPECIALIZED AND DIVERSE MAMMALS

With over 400 species, the insectivores are the third largest order of mammals. They are largely confined to the northern temperate zones of North America, Canada, Europe, and Russia, but also occur in Africa, southern Asia, and northern South America. Most insectivores have primitive brains and depend more on sense of smell than on vision; they usually have specialized dentitions; their limbs are almost always unspecialized, and they have generalized quadrupedal locomotion.

The taxonomy of this diverse group has been, and still is, the subject of much controversy. In the past, the group has often been treated as a common dumping ground for those species that do not share clear affinities with other groups or that did not, when classified, merit being placed in their own distinctive order. Probably no other group of mammals has such a clouded history as those that have developed the specialization of insectivory, or insect eating. Fossil evidence now indicates that the most primitive placental mammals were insectivores, and, since their descendants have retained dentitions that are still adapted to an insect diet, taxonomists have tended to group all types of this primitive form together with all modern descendants of its ancient lineage. Today, however, it has been recognized that the morphological differences between different forms are sufficient to warrant a revision of this classification, and three separate orders of insect-eating mammals are now recognized: the Insectivora, which include such familiar species as hedgehogs, moles, tenrecs, and shrews; the order Scandentia or tree shrews of Asia; and the order Macroscelididae or African elephant-shrews.

The order Insectivora is rich in examples of convergent evolution (where animals not closely related have adapted behaviorally or morphologically to fit the demands of a specific habitat or way of life). As an example, three diverse and geographically separated species—the Pyrenean desman *Galemys pyrenaicus,* the Mount Nimba otter shrew *Micropotamogale lamottei,* and the Madagascan aquatic tenrec *Limnogale mergulus*—have evolved in total isolation from each other, but, probably in the absence of aquatic predators and in response to the availability of an unexploited niche, each has evolved toward an aquatic way of life. The result is that they all share some physical modifications: a streamlined body shape with a reduction of external appendages, partially webbed feet, a long tail that can act as a rudder, a specialized breathing apparatus, a dense, waterproof coat, and specialized sensory mechanisms for detecting prey underwater. Other, more general, examples of convergent evolution within this order include exploitation of the fossorial (burrowing) niche by both European moles and African golden moles.

SOLENODONS

Among the largest of the living insectivores, the secretive solenodons (family Solenodontidae) are

P. Morris/Ardea London

▼► *Moonrats (top right) are not rodents but close relatives of hedgehogs. Their name refers to the naked, rat-like tail. Least specialized of the insectivores, they feed on a variety of prey. The tenrec family, largely confined to Madagascar, has undergone an amazing evolutionary radiation from shrew-like creatures into forms resembling hedgehogs, moles, and otters. The skunk-like coloration of the streaked tenrec (below, left) advertises that it is unpalatable. Like their relatives, the typical moles, desmans have small eyes and a sensitive snout. The Pyrenean desman (below, right) lives in mountain streams and feeds on bottom-dwelling invertebrates. These illustrations are not to scale; in fact, the moonrat (at about 30 centimeters or 12 inches) is roughly twice as long as the streaked tenrec, which in turn is about double the size of the Pyrenean desman.*

currently confined to Cuba and Hispaniola in the Greater Antilles. Represented today by just two species—*Solenodon paradoxus* and *S. cubanus*—these relict species are now unfortunately recognized as being among the most highly endangered species on earth.

Before the arrival of humans on these islands, the solenodon was probably one of the top carnivores of the ecosystem, although it, in turn, would probably have been prey to large owls and boas. The arrival of humans, however, almost certainly spelled doom for these predominantly nocturnal creatures. Dogs and rats soon invaded the islands from fishing vessels, and, in an attempt to control the rats, which were causing widespread destruction and disease, the mongoose was introduced at a later stage. Each of these exotic introductions represented new, highly intelligent predators and competitors for the solenodons. In fact, there is evidence that the now extinct genus of shrews *Nesophontes* was once widespread throughout the West Indies, and its disappearance coincided with the arrival of the Spaniards. Today,

the few remaining solenodons appear to be destined to suffer the same fate.

As in most nocturnal insectivores, brain size is relatively small, but the sense of touch is highly developed. When feeding, the solenodon's extended and highly sensitive snout is operated like a tentacle, probing for prey in cracks and crevices while its strong claws are used to dig or break up rotting vegetation. Once the prey is located, it is quickly bitten, and a small amount of toxic saliva is injected into the victim, causing paralysis. The solenodon probably uses this technique more often when capturing prey larger than insects, such as frogs, lizards, or small birds.

The natural history of solenodons is poorly known, and the animals have not been studied in great detail in the wild or in captivity. Their life history is characterized by a relatively long life-span and low reproductive potential (only one offspring is produced each 12 to 18 months). The single young is born in a specially created nesting burrow and, once weaned, accompanies the mother on her nightly foraging trips. Solenodons

S.C. Bisserot

▲ *The resemblance of the hedgehog tenrec to "true" hedgehogs is the result of convergent evolution in two distinct families of insectivores. Tenrecs retain some primitive mammalian features: males lack a scrotum and (like monotremes and marsupials) females have a cloaca.*

are the only insectivores to practise teat transport, a technique that probably facilitates the rapid movement of young from one burrow to another if the animal is disturbed. All other insectivores carry their young in their mouths.

Solenodons appear to be strictly solitary in their habits, the mother–offspring tie being the only period of extended social behavior. Being nocturnal, solenodons rely on auditory and olfactory signals to locate their prey and also to monitor the presence of other animals.

Solenodons inhabit well-established open forests that generally have a considerable amount of scrub or ground cover. Unfortunately, almost nothing is known about their social or spatial behavior in the wild.

TENRECS AND OTTER SHREWS
Resembling hedgehogs in their general appearance and, in part, in their way of life, the tenrecs of

Madagascar and the otter shrews of Central Africa provide a striking example of convergent evolution and diversification. Fossil records indicate that members of the family Tenrecidae were already well established within the general African fauna 25 million years ago. Today, however, the only surviving members of this ancient lineage on the African mainland are the three species of otter shrew, which are confined to small rivers in the tropical forest belt of West and Central Africa.

The tenrecs were among the first mammals to arrive on Madagascar after its separation from mainland Africa around 150 million years ago. Arriving on an island with no worthy competitors, the founder species radiated into a wide variety of ecological niches. The radiation of tenrecs is one of the classic examples of radiation, with representatives of fossorial, terrestrial, semi-aquatic, and even semi-arboreal tendencies. In general, tenrecs are small animals—the largest,

Tenrec ecaudatus, weighs 1 kilogram (2 pounds), and the smallest, *Geogale aurita,* a termite-feeding specialist, weighs less than 10 grams (⅖ ounce). They are either crepuscular (active at twilight) or fully nocturnal in habit, and most of them are solitary. An exception to this rule are members of the species *Hemicentetes semispinosus,* which appear to form colonies during the breeding season.

GOLDEN MOLES

Golden moles—so called because of the iridescent bronze glint on their fur—are strictly African in their distribution. Specialized for a fossorial life, the 18 species of the family Chrysochloridae display physical and behavioral adaptations to a wide range of terrestrial habitats.

The golden mole's streamlined body is covered in a relatively coarse, backward-facing fur, which is moisture repellant, superimposed upon a much softer, denser undercoat, which provides insulation. The eyes are minuscule and probably rudimentary, being covered with a hairy skin. The ear openings are covered in dense fur, and the nose is protected by a tough leathery pad, which may be used to move loose soil when digging.

The stocky, thickset body gives a semblance of strength and, indeed, one early naturalist remarked on how a captive golden mole could exert a force equal to 150 times the animal's own weight. The forelimbs are extremely powerful and are armed with four short digging claws. When digging, the hindfeet are firmly braced against the tunnel wall and the soil at the digging face is loosened by repeated, sharp, downward strokes of alternate forelimbs. Displaced soil is either packed against the side or roof of the tunnel using the broadened head and shoulders, to form surface ridges, or is kicked backward along the tunnel for some distance from where it is pushed to the surface by the head, to form surface molehills. The digging behavior of golden moles, which use a combined action of head, feet, and body, differs significantly from that of European and North American moles, which use only their feet.

Golden moles are solitary animals that spend more time on burrowing than on any other single activity. So great is this need that some species spend up to 75 percent of their active time engaged in burrowing. The style and size of the burrow system depends upon the species and its habitat. Mountain-dwelling and plain-dwelling species, such as the giant golden mole *Chrysospalax trevelyani* or the Hottentot golden mole *Amblysomus hottentotus,* build a semipermanent series of tunnels, which may descend to almost 1 meter (3 feet) in depth. These, and similar species, trap and eat food within the tunnels as European and North American moles do. Food is eaten immediately; it is not cached or stored. Desert-dwelling species, however, such as Grant's golden mole *Eremitalpa granti,* do not form permanent

tunnels, but instead push their way through the upper reaches of their sand-dune ecosystem almost in a swimming action, locating prey by touch and perhaps hearing. Desert moles may occasionally hunt on the surface of the sand, where they trap legless lizards or insects.

HEDGEHOGS AND MOONRATS

Probably one of the most familiar insectivores, which in several ways resembles a diminutive solenodon, is the European hedgehog *Erinaceus europaeus.* Hedgehogs belong to the family Erinaceidae, which has representative members throughout Europe, Africa, and Asia. This group has about 21 species, some of which (for example, the stocky European hedgehog and the desert hedgehogs of Asia and North Africa) bear spines, while others, such as the intriguingly named moonrats or gymnures of Southeast Asia, are covered with coarse hair and often have long tails.

The most distinctive characteristic of the hedgehog is its dense coat of spines—an adult will have as many as 5,000 needle-sharp spines 2 to 3 centimeters (¾ to 1 inch) in length. The spines are actually modified hairs, each of which is filled with multiple air chambers and strengthening ridges that run down the inside wall of each tube. The spines cover the entire dorsal surface of the body, while the ventral surface is covered with a toughened skin from which protrudes a covering of coarse hair. Normally the spines are laid flat against the body but, when erect (for example, if the animal is threatened) they stick out at a variety of angles, overlapping and supporting each other to create a truly formidable defense system. As if this were not sufficient protection, hedgehogs are also able to curl into a tight ball, thereby fully concealing and protecting the softer underparts of the body. Confronted with such defense, few predators would even attempt to attack.

▼ *The golden moles of Africa look remarkably like the marsupial mole of Australia—a fascinating example of convergent evolution. The eyes are tiny and probably useless; there is no external ear; the snout is protected by a horny shield; and the forelimbs are powerful, spade-like structures.*

G.J. Broekhuysen/Ardea London

Hedgehogs are solitary animals and, depending upon the species and climate, have quite different breeding patterns. Arid-dwelling species, such as desert *Paraechinus* species and long-eared *Hemiechinus* species, breed only once a year. In temperate zones two litters are known to occur regularly. In tropical zones, where food availability and climate (two of the major determinants of the timing and duration of the breeding season) are more predictable, hedgehogs can breed all year.

The unavailability of or difficulty in locating prey is a serious consideration for all insectivores, because of their high energy demands. To combat the need to remain active when prevailing environmental conditions are not favorable, many species have developed the ability to undergo a period of dormancy (torpor or hibernation), during which the body temperature is allowed to decrease to a level close to that of the surrounding air. For example, oxygen requirements for a hedgehog may decline from an average (normal) level of 500 milliliters per kilogram per hour to about 10 milliliters per kilogram per hour. This strategy enables species that are experiencing some degree of crisis, for example food and temperature extremes, to dramatically reduce their energy expenditure, which effectively enables the animals to survive longer on fewer reserves.

Hibernation is not a species trait but is dictated by environmental conditions. Tropical hedgehogs will not normally hibernate but, if artificially exposed to low food levels or low ambient temperatures, these species also exhibit the ability to hibernate. The European hedgehog, which undergoes a seasonal hibernation in its native land, forgoes this behavior in New Zealand, where it was introduced at the beginning of this century.

Temperate-dwelling hedgehogs are commonly found in deciduous woodlands, arable land surrounded by hedgerows, and urban gardens. During the summer months, a simple nest is

▼ *Like other hedgehogs, the long-eared hedgehog has a head and upper body covered with spines. These hedgehogs are arid zone dwellers; their long ears act as heat radiators. They and the desert dwelling species breed only once a year.*

Eyal Bartov/Oxford Scientific Films

constructed from leaves and grasses. and is usually placed among the undergrowth at the base of a tree, rather than underground. Such nests are rapidly constructed and may be abandoned after just a few days in preference to another. In winter, more care is taken when choosing the nest site, since this must be well insulated against the cold.

When active, hedgehogs are constantly on the lookout for prey, and, in the course of a single night's foraging, some animals have been observed to travel over 3 kilometers (2 miles). Their favorite food is earthworms, although they will eat most ground-dwelling invertebrates, as well as seeds and fruit. In Europe, many people actually attempt to attract hedgehogs to their gardens—their appetite

for slugs and chafer beetles making them the gardener's friend—by leaving out bowls of bread and milk, which the animals do appear to relish.

From what little is known about the Asian gymnures they, like hedgehogs, are predominantly solitary and do not appear to defend a fixed territory. Being nocturnal, their primary mode of communication appears to be olfaction (the sense of smell), a feature that is keenly developed in the greater moonrat *Echinosorex gymnura* and in the lesser moonrat *Hylomys suillus,* both of which have well-developed anal scent glands that exude what is, to the human sense of smell, a foul odor. When placed in strategic parts of the animal's domain, this substance indicates that the area is already

occupied and may possibly identify the sex, age (and hence the dominance status), or stage of sexual maturity of the resident animal.

Moonrats are among the largest of the insectivores, occasionally weighing up to 2 kilograms (4½ pounds). They look quite ferocious with their coarse, shaggy coat and impressive, open-mouthed, threatening gestures. Moonrats inhabit lowland areas, including mangrove swamps, rubber plantations, and primary and secondary forests. Largely nocturnal and terrestrial, they rest during the day in hollow logs, under the roots of trees, or in empty holes. They appear to prefer wet areas and often enter water to hunt for insects, frogs, fish, crustaceans, and mollusks.

▲ *Although hedgehogs are solitary animals, the partially weaned young accompany their mother to forage at night for insects and worms.*

A. & E. Bomford/Ardea London

▶ *Like most members of its family, the common shrew lives on the forest floor, actively feeding on insects, worms, and other invertebrates. Most shrews are so small that they must feed continuously to maintain body temperature. Without food, a shrew can die of starvation in as little as four hours.*

▼ *The Eurasian water shrew feeds mainly on bottom-living invertebrates but also takes frogs almost as large as itself. Such prey is killed by venomous saliva which is injected when the teeth of the shrew penetrate its flesh.*

SHREWS

Of all insectivores, the diminutive shrews are the most successful single group. They are successful both in terms of their evolutionary radiation (almost 250 species) and also in their geographical distribution, which includes all of North and Central America, Europe, Asia, and most of Africa. Shrews are small; the smallest species, *Suncus etruscus*, weighing only 2 grams (7/100 ounce), is the world's smallest mammal. Shrews are secretive mammals, and are characterized by long, pointed noses, relatively large ears, tiny eyes, and dense velvety fur.

The family Soricidae, to which shrews belong, is divided into two groups: the so-called red-toothed shrews, which have pigmented teeth and are northern European in origin, and the white-toothed shrews, which originated in Asia and North Africa and have now colonized much of mainland Europe.

Shrews are very active animals and their hectic pace of life means that they must consume disproportionately large amounts of food for their body size. Recent research has shown that many species of shrew have a much higher metabolic rate than rodents of comparable size. To cope with such constant energy demands, shrews tend to live in

Dwight R. Kuhn

habitats that are highly productive. To survive, they must feed every two or three hours, and they have been recorded to consume 70 to 130 percent of their own body weight each day. Not surprisingly, therefore, shrews are generally highly opportunistic feeders, taking a wide range of invertebrate prey, supplemented by occasional carrion, fruit, seeds, and other plant material. Digestion in shrews is fairly rapid, and the gut may be emptied in under three hours. In temperate climates, at least, most species are highly territorial, aggressively defending a fixed piece of terrain against all intruding shrews, thereby ensuring that they will have undisturbed access to the maximum amount of prey.

Shrews occupy a wide range of feeding habitats, exploiting aquatic, terrestrial, and fossorial niches. Some fossorial species make use of existing tunnels dug by moles and rodents, but some red-toothed shrew species, for example the short-tailed shrew *Blarina brevicauda*, have adapted toward a truly fossorial way of life, and their body form is similar to that of the true moles. Prey, detected mainly by hearing and, to a lesser extent, by smell, is rapidly seized and killed. At least two species—the American short-tailed shrew *Blarina brevicauda* and the Eurasian water shrew *Neomys fodiens*—are known to inject their prey with venomous saliva (as solenodons do), which may be particularly useful when dealing with large prey such as frogs or fish. If there is an abundance of prey, part of a catch may be cached or stored for later use. These caches are usually defended quite vigorously against other shrews.

Shrews are essentially solitary animals and it is only at breeding times that they will approach one another in an amicable manner. Like moles, shrews are sexually receptive for a very brief period. In some shrews this may be seasonal, but others may breed at any time of the year. In *Suncus*, the house shrew, for example, although a female may not be reproductively active, a male may still mount her and bite the back of her neck until she becomes receptive, when she emits a characteristic chirp. The aggressive advances and odor-marking of the male thus appear to bring the female into heat, when mating and fertilization may take place.

Young shrews are born after a three-week gestation period and remain confined for some time in a specially prepared nest chamber, which is usually underground for protection. The mother is highly attentive and communicates with the litter of three to eight (depending on species) by making faint squeaks, barely audible to the human ear. If the young stray from the nest the mother will retrieve them, carrying them in her mouth back to the nest site.

At three weeks, the young are encouraged to leave the nest and accompany the mother on short foraging trips. A peculiar habit, characteristic of the white-toothed shrews, is caravanning, where the

Stephen Dalton/NHPA

young, when foraging, actually line up and grab a tuft of hair on the rump of the preceding animal—the one at the front holding onto the mother in the same way. This peculiar behavior is first and foremost a mechanism for the young to avoid a predator.

As with moles, if the young persist in staying at the maternal nest they will be driven away by aggressive gestures. By the time the litter has dispersed, the female will probably have mated again. In this way, adult breeding females can give birth to two or even three litters a year, although the adults themselves will not survive the winter period. Compared with other similar-sized mammals, shrews have a very short life-span of 9 to 12 months on average, although all can live longer in captivity, the record being four years for a greater white-toothed shrew *Crocidura russula*.

▲ The dense fur of the water shrew is strongly water-repellent. When the animal dives, its body is surrounded by air trapped between the hairs. Where streams are polluted by detergent, the fur of water shrews becomes wet and they die of cold.

MOLES AND DESMANS

The family Talpidae includes the moles, shrew moles, and desmans, all of which are confined to North America and Eurasia. These predominantly burrowing insectivores (42 species in 17 genera) are highly secretive and because of their way of life have, in general, been poorly studied. The species that has, to date, received most attention from naturalists and biologists alike is the European mole *Talpa europaea,* whose way of life and behavior are probably quite similar to many of the other species within this family.

Moles are highly specialized for a subterranean, fossorial way of life. Their broad, spade-like forelimbs, which have developed as powerful digging organs, are attached to muscular shoulders and a deep chestbone. The skin on the chest is thicker than elsewhere on the body as this region supports the bulk of the mole's weight when it digs or sleeps. Behind the enormous shoulders the body is almost cylindrical, tapering slightly to narrow hips, with short sturdy hindlimbs (which are not especially adapted for digging), and a short, club-shaped tail, which is usually carried erect. In most species, both pairs of limbs have an extra bone that increases the surface area of the paws, for extra support in the hindlimbs, and for moving earth with the forelimbs. The elongated head tapers to a hairless, fleshy, pink snout that is highly sensory. In the North American star-nosed mole *Condylura cristata,* this organ bears 22 tentacles each of which bears thousands of sensory organs.

The function of a mole's burrow is often misunderstood. Moles do not dig constantly or specifically for food. Instead the tunnel system, which is the permanent habitation of the resident

animal, acts as a food trap constantly collecting invertebrate prey such as earthworms and insect larvae. As they move through the soil column invertebrates fall into the animal's burrow and often do not escape before being detected by the vigilant, patrolling resident mole. Once prey is detected, it is rapidly seized and, in the case of an earthworm, decapitated. The worm is then pulled forward through the claws on the forefeet, thereby squeezing out any grit and sand from the worm's body that would otherwise cause severe tooth wear—one of the common causes of death in moles. If a mole detects a sudden abundance of prey, it will attempt to capture as many animals as possible, storing these in a centralized cache, which will usually be well defended. This cache, often located close to the mole's single nest, is packed into the soil so that the eathworms remain

alive but generally inactive for several months. Thus, if an animal experiences a period of food shortage it can easily raid this larder instead of using essential body reserves to search for scarce prey. In selecting such prey for the store, moles appear to be highly selective, generally choosing only the largest prey available.

Tunnel construction and maintenance occupy much of a mole's active time. A mole digs actively throughout the year, although once it has established its burrow system, there may be little evidence above ground of the mole's presence. Moles construct a complex system of burrows, which are usually multi-tiered. When a mole begins to excavate a tunnel system, it usually makes an initial, relatively straight, exploratory tunnel for up to 20 meters (22 yards) before adding any side branches. This is presumably an attempt to locate neighboring animals, while at the same time forming a food trap for later use. The tunnels are later lengthened and many more are formed beneath these preliminary burrows. This tiered-tunnel system can result in the burrows of one animal overlying those of its neighbors without them actually being joined together. In an established population, however, many tunnels between neighboring animals are connected.

Moles have a keen sense of orientation and often construct their tunnels in exactly the same place every year. In permanent pastures, existing tunnels may be used by many generation of moles. Some animals may be evicted from their own tunnels by the invasion of a stronger animal and, on such occasions, the loser will have to go away and establish a new tunnel system. These "master engineers" are highly familiar with each part of their own territory and are suspicious of any changes to a tunnel, which makes them difficult to capture. If, for example, the normal route to the nest or feeding area is blocked off, a mole will dig either around or under the obstacle, rejoining the original tunnel with minimum digging.

Our knowledge of the sensory world of moles is very limited. They are among the exclusively fossorial species; the eyes are small and concealed by dense fur or, as in the blind mole *Talpa caeca,* covered by skin. Shrew moles, however, forage not only in tunnels beneath the ground but also above ground among leaf litter. Although they may have a keener sense of vision than other species, they are still probably only able to perceive shadows rather than rely heavily on vision for detecting prey or for purposes of orientation. The apparent absence of ears on almost all species is due to the lack of external ear flaps and the covering of thick fur over the ear opening. It has, however, been suggested that ultrasonics may be an important means of communication among fossorial and nocturnal species. But of all the sensory means olfaction appears to be the most important medium—a fact supported by the elaborate nasal region of many

◀ *"True" moles of the family Talpidae are found over much of Eurasia and North America. Their eyes are tiny and useless for vision; there is no external ear but their hearing is acute. The long snout is extremely sensitive and the sense of smell is well developed.*

Dwight R. Kuhn

▲ The sensory tentacles on the snout of the star-nosed mole are unique among mammals. This species burrows in swampland but also spends much time above ground and even enters the water in search of prey, paddling with its spade-like forelimbs.

their mother for warmth. The young are fed entirely on milk for the first month, during which they rapidly gain weight. Juveniles remain in the nest until they are about five weeks old, at which time they begin to make short exploratory forays in the immediate vicinity of the nest chamber. Shortly thereafter they begin to accompany their mother on more extensive explorations of the burrow system and may disperse from there of their own accord; those that do not leave will soon be evicted by the mother.

A very different way of life is exhibited by the desmans, of which there are two species. The Pyrenean desman *Galemys pyrenaicus* is confined to permanent, fast-flowing streams of the Pyrenees mountain range and parts of northern Iberia; the Russian desman *Desmana moschata* is found only in the slower moving waters and lakes of the western and central Soviet Union. Just as moles are superbly adapted for a fossorial way of life, so too are the desmans for water. The streamlined body of the Pyrenean desman enables it to glide rapidly through the water, propelled by powerful webbed hindlimbs and steered, to some extent, by a long, broad tail. For any animal living in the snow-fed mountain streams, feeding and retaining body heat are top priorities. Unlike hedgehogs or tenrecs, desmans do not undergo periods of hibernation or torpor and must, therefore, live in optimum habitats to ensure their survival during the winter months when prey is most scarce.

Desmans feed on the larvae of aquatic insects such as the stone fly and caddis fly, as well as on small crustaceans, which they locate by probing their proboscis-like snouts beneath small rocks and by clearing away debris from the stream bed with their sharp elongated claws. Prey is consumed at the surface where, following each dive, a rigorous body grooming is carried out. This is an essential activity as it ensures that the fur is not only kept clean and in good condition but also maintains its water-repellent properties by spreading oil all over the body from sebaceous glands.

Desmans construct their nests in the banks of streams. The Russian desman actually excavates a complex burrow, which it may share with other desmans, while the smaller Pyrenean species occupies a strictly solitary nest, usually created by enlarging an already existing tunnel or crevice. Nests are composed of leaves and dried grasses and are always located above the water level.

Little is known about the breeding behavior of desmans. In the Pyrenean desman, mating takes place in spring (March to April), and, as these animals usually form a stable pair bond, competition for mates by solitary males is often quite severe. At this time of year, an interesting phenomenon occurs for each pair of animals: males become far more protective, spending most of their active time at the upper and lower reaches of their riverine territory. Energy is thus spent on

species, together with the battalion of sensory organs stored within this area.

The brief breeding season is a frantic period for moles, as females are receptive for only 24 to 48 hours. During this time males usually abandon their normal pattern of behavior and activity, spending large amounts of time and energy in locating potential mates. Mating takes place within the female's burrow system and this is the one period of non-aggressiveness between the sexes. The young, with an average of three to the litter, are born in the nest four weeks later. Weighing less than 4 grams (⅐ ounce), the pink, naked infants cannot control their body temperature and rely on

David Thompson/Oxford Scientific Films

protecting the feeding resources of that territory and, more importantly, the female. Females, in contrast, spend most of their active time feeding, surveying for a suitable nest site, and gathering nesting materials.

Young are born after a gestation period of about four weeks and are cared for solely by the female. Juveniles first leave the nest at about seven weeks, at which stage they are already proficient swimmers. Juveniles remain within the parents' territory until they are about two and a half months old, at which stage they leave in order to secure a mate and breeding territory for the coming year.

CONSERVATION

From a conservation standpoint, many species of insectivores are now severely threatened. Unlike many other mammals, this threat does not come from direct human exploitation of the animals but indirectly through such activities as deforestation, introduction of exotic species to the ecosystem, pollution, and other activities related to habitat destruction. These activities almost invariably arise as a consequence of human exploitation of or interference with the balance of the ecosystem.

Probably one of the most severely threatened genera is *Solenodon*. No recent estimates of the number of surviving animals exist, but it is known that their habitat—a mixture of forest and rocky outcrops—is rapidly diminishing.

In fact, habitat destruction appears to be the single greatest threat to insect-eating organisms. It is a major problem for the giant golden mole of Africa, as the relict forests and open grasslands where this localized species occurs are being rapidly converted into second-class grazing and arable pastures. Aquatic insectivores too are facing serious threats to their survival. In Europe, the Pyrenean desman is threatened by the construction of hydroelectric dams in its mountainous refuge; the endemic otter shrews of Africa and the aquatic tenrec of Madagascar are threatened by deforestation and increased siltation of rivers.

R. DAVID STONE

▲ The European mole feeds mostly on the earthworms that enter its tunnels. When earthworms are plentiful, it bites the heads off some and stores them, still alive, in a "larder" close to its nest.

TREE SHREWS

Tree shrews, of the order Scandentia, are for the most part arboreal mammals, although some species are also found living at ground level. Primarily inhabitants of tropical rainforests, these fascinating but relatively unknown animals are believed to be one of the most primitive forms of placental mammals and, as such, representative of the ancestral stock from which present-day mammals have evolved.

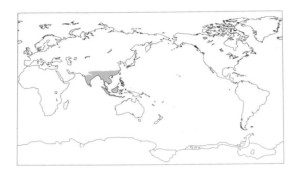

SHREWS LIKE SQUIRRELS

The 19 living species of tree shrews are all confined to eastern India and Southeast Asia. The order contains a single family, the Tupaiidae, which is further subdivided into two subfamilies, the Ptilocercinae and the Tupaiinae. Superficially, tree shrews resemble small tree squirrels—the genus name *Tupaia* is derived from the Malay *tupai,* meaning "squirrel-like animal"—but they differ in both anatomy and behavior. Tree shrews are relatively small mammals, averaging 70 to 100 grams (2½ to 3½ ounces) weight. Generally a russet brown color, they have a long, pointed muzzle with 38 sharp, pointed teeth, and 5 clawed digits on each foot. Tail length usually exceeds body length, but the tail is not prehensile.

FEATHER-TAILED TREE SHREW

Within the subfamily Ptilocercinae is just a single species, the feather-tailed or pen-tailed tree shrew *Ptilocercus lowii*, which is nocturnal and is found only on the Malay Peninsula and Borneo. Weighing less than 80 grams (3 ounces) and measuring 30 to 35 centimeters (12 to 14 inches) from head to tail, this strange-looking gray-brown animal has large ears and long facial hairs. Its tail, uniquely among tree shrews, is covered in scales, except for the tip, which has long white hairs growing out of opposite sides, giving a feather-like appearance. The tail is quite long and serves as an organ of balance, support, and touch: when an animal is awake, the tail twitches continuously.

Feather-tailed tree shrews are arboreal and have a number of distinctive adaptations to this niche: the hands and feet are relatively larger than in other tree shrews, and the toes can be spread more widely, giving a better grip when climbing and also permitting the shrew to grasp insects with only one hand. The digits are flexed so that the claws are always in contact with the surface of the branch,

and the footpads are also larger and softer than in other species. This species appears to be sociable, living in small groups and often sharing a common sleeping nest. It feeds predominantly on a mixture of fruit and insects, particularly cockroaches, beetles, ants, and termites.

OTHER TREE SHREWS

The remaining 18 species of tree shrew are classified within the subfamily Tupaiinae and are distributed throughout eastern India, Southeast Asia, and various parts of the Malaysian archipelago. Most species are omnivorous and, depending on the species and habitat, are either arboreal or terrestrial in habit. They range from lowland forest through secondary and primary rainforests and even to montane habitats where one species, the mountain tree shrew *Tupaia montana,* is found. These diurnal tree shrews differ from their predominantly nocturnal feather-tailed relatives in their better-developed vision and greater brain development. These characteristics separate the tree shrews from other insectivores, such as hedgehogs, shrews, and moles.

Tree shrews are highly active, nervous, inquisitive, and generally aggressive animals. They are very fond of water and often bathe in water-filled hollows of trees

The common tree shrew *Tupaia glis* lives in permanent pairs. Although the partners of each pair exhibit solitary daily ranging behaviors, they occupy the same territory, which they defend vigorously against other members of their species of the same sex. As a rule, the territories of different pairs overlap only slightly. Young animals apparently live with their parents, thus forming family groups. After attaining sexual maturity, young animals are forced to leave, and they adopt a nomadic existence for several months until they find a suitable mate and establish their own territory. Tree shrews may live for between two and three years in the wild and display a strong degree of fidelity to their mates.

LIVING IN THE TERRITORY

Scent marking appears to be a very important part of the daily routine for all species of tree shrew. Scent marks generally transmit information about the animal that deposited them and may be dispersed in either a passive manner, for example general body odor that exudes through skin pores, or in a highly selective, active fashion, whereby specific scents are deliberately deposited at a strategic part of the animal's territory, with the intention that these will be detected and interpreted by similar species. Thus, if an animal is able to detect odor and differentiate the many components of a scent mark, it may be able to gain information about the sex and social status of the depositor, and how recently the mark had been made. Tree shrews regularly distribute secretions

Rod Williams/Bruce Coleman Ltd

from specialized scent glands to new objects within the territory, and more often at specific parts of the territory, to indicate to neighboring animals that that site is already occupied and is being defended. Such boundary marking sites are often used by several neighboring animals and thus serve as communication points, providing a wealth of local information to the resident animals.

In addition to saturating the territory with odor, pair-living animals will daily cover each other with their own respective odors. Parents will also mark juveniles once they begin to leave the nest; captive females have been observed to mark their young over one hundred times in a single day.

The breeding behavior of free-living tree shrews has not been investigated in any detail, but observation of captive animals has revealed some intriguing facts. There is no indication of a fixed breeding season: breeding may occur throughout the year. A few days before the female is due to give birth, the male prepares a "maternal nest" of leaves and will then leave and not return to the nest site until the young are at least one month old. In the forest, nest sites are usually located in holes in fallen trees, hollow bamboos, or similar sites.

After a gestation period of 40 to 50 days, the naked young are born in the nest and are suckled immediately. Litter size varies according to species, but an average of three young are born. The mother assumes all responsibilities for feeding and defending the young until they have been weaned. If the young are disturbed at the nest, they make a very loud, abrupt sound and simultaneously thrust out all four legs, loudly rustling the leaves in the nest and creating a startling display that may discourage a predator. The main natural predators are birds of prey, small carnivores, and snakes.

R. DAVID STONE

▲ *In the past, tree shrews have been classified as insectivores and as primates. The reality appears to be that they are very unspecialized mammals that share some features with these two groups. The large tree shrew, shown here, is terrestrial and has a long snout and strong digging claws; the more arboreal tree shrews have shorter faces and more slender claws.*

FLYING LEMURS

ORDER DERMOPTERA
• 1 family • 1 genus • 2 species

SIZE

Malayan flying lemur
*Cynocephalus variegatus**
Head–body length: 34–42 cm
(13–17 in)
Tail length: 22–27 cm (9–11 in)
Weight: 1–1.75 kg (2–4 lb)

* The Philippine flying lemur
Cynocephalus volans is slightly
smaller.

CONSERVATION WATCH
! The Philippine flying lemur
Cynocephalus volans is listed as
vulnerable.

▶ *Flying lemurs feed on leaves, shoots
and buds of rainforest trees. The
Malayan flying lemur (shown here) is a
nuisance in plantations because of its
habit of eating coconut flowers.*

▼ *Flying lemurs are among the largest of
the gliding mammals. Unlike the others,
they have a membrane between the tip
of the tail and the ankles, which can be
folded into a soft pouch for carrying the
young. The young are also carried
clinging to the mother's belly.*

F lying lemurs, or colugos, belong to the very small order Dermoptera, which
includes the two living species, genus *Cynocephalus,* in the family
Cynocephalidae from Southeast Asia. The common name "flying lemurs" is
misleading since they do not fly but glide, and they are not lemurs but the surviving
representatives of a large evolutionary lineage possibly related to the ancestral
primates. Indeed many ancient dermopterans were until recently considered to be
ancestors of the primates.

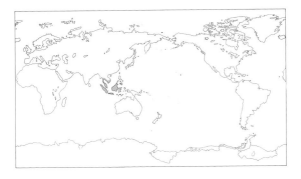

MAMMALS THAT GLIDE
Flying lemurs are cat-sized, with grayish or reddish
brown backs, which are usually speckled and
mottled with black and grayish white markings.
The arrangement of the unusual and distinctive
incisor teeth is similar to that of herbivorous
mammals such as cattle or deer. The upper incisors
are at the sides of the jaw, leaving a gap at the front,
and do not oppose the forwardly directed lower
incisors. These are broad and comb-like, with as
many as 20 "tines" arising from one root; they may
provide a scraping action for abrading and
straining food, and grooming and cleaning the fur.
The feet are furnished with strong claws.

The order name Dermoptera, meaning skin
wing, aptly describes their most striking
feature—the gliding membrane, or patagium, that
stretches from the neck to the finger tips, along the
sides of the body, between the toes, and joins the
legs and tail. The gliding membrane enables flying
lemurs to glide up to a recorded 136 meters (150
yards). The Philippine flying lemur *Cynocephalus
volans* is slightly smaller than the Malayan flying
lemur *Cynocephalus variegatus,* and the fur on its
back is darker and less spotted.

Flying lemurs live in primary and secondary
tropical forest, and in rubber and coconut
plantations, where their spectacular gliding ability
enables them to move easily from tree to tree
while searching for food in the canopy. They are
nocturnal, spending the day in a tree hollow or
hanging beneath a branch and moving off at night
through the trees to a preferred feeding area.
Flying lemurs climb clumsily in a lurching fashion,
gripping the tree with outspread limbs, and
moving first the forefeet and then the hindfeet.
Gliding from the upper part of one tree to a lower
point on another, they climb again to make
another glide to a further tree, so progressing
through the forest, often along a regular route and
with several animals following each other. Flying
lemurs are herbivores, subsisting on leaves,
shoots, buds, and perhaps flowers and fruit. They
pull twigs and small branches within reach with
the forefeet, stripping the leaves with the lower
incisors and long, strong tongue. Their digestive
system is specialized for this vegetarian diet, with
a large cecum and long, convoluted intestine.

Gestation takes 60 days. Usually a single young
is born, although occasionally twins are
produced. Newborn flying lemurs are relatively
undeveloped and, until weaned, are carried
clinging to the belly of the mother, who can also
fold the gliding membrane near the tail into a soft,
warm pouch for this purpose.

ORIGINAL TEXT BY J. E. HILL;
REVISED BY GEORGE MCKAY

D. &W. Ward/Oxford Scientific Films

BATS

ORDER CHIROPTERA
• 18 families • *c.* 180 genera
• *c.* 925 species

SMALLEST & LARGEST

Hog-nosed bat *Craseonycteris thonglongyai*
Forearm length: 22.5–26 mm (⁹⁄₁₀–1 in)
Wing span: 15 cm (6 in)
Weight: 1.5–2 g (⁵⁄₁₀₀–⁷⁄₁₀₀ oz)

Large flying fox *Pteropus vampyrus*
Forearm length: 22 cm (8½ in)
Wing span: 2 m (79 in)
Weight: up to 1.2 kg (2⅗ lb)

CONSERVATION WATCH
!!! 25 species are critically endangered, including: Bulmer's fruit bat *Aproteles bulmerae*; cusp-toothed fruit bat *Pteralopex atrata*; Rodrigues flying fox *Pteropus rodricensis*; Seychelles sheath-tailed bat *Coleura seychellensis*; Wroughton's free-tailed bat *Otomops wroughtoni*.
!! 32 species are listed as endangered, including the big long-nosed bat *Leptonycteris nivalis* and gray bat *Myotis grisescens*.
! 175 species are listed as vulnerable.

Bats occur almost worldwide, except in the Antarctic, the colder area north of the Arctic Circle, and a few isolated oceanic islands, although more species are concentrated in the tropics and subtropics than in the temperate zones. Some 977 species are currently recognized, second only in number to the rodents. Bats are the only mammals capable of true flight, and their wings are their most obvious feature. To navigate, avoid obstacles, and feed at night, when most are active, bats have developed superb hearing and the use of high frequency echolocation.

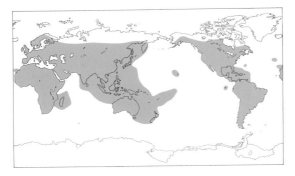

MAMMALS THAT FLY

The order Chiroptera is divided into two suborders: the Megachiroptera, or Old World fruit bats, that include only 1 family with about 41 genera and 163 species, and the Microchiroptera, basically insect-eating bats, with 17 families, about 146 genera and 814 species. The terms Megachiroptera (large bats) and Microchiroptera (small bats) are not totally definitive—many species in the suborder Microchiroptera are larger than the smaller species in the suborder Megachiroptera. The numbers of genera and species are not immutable: they alter from time to time as classification changes or new bats are described.

Little is known of the origins of bats. The earliest known fossils come from the Eocene (55 million years ago) of West Germany and North America. These fossils are similar to modern bats and it is suspected that bats originated much earlier, perhaps 70 to 100 million years ago, possibly from small quadrupedal, arboreal, and insectivorous ancestors that developed gliding membranes and may have been related to ancestral insectivores or dermopterans.

The wings of bats are their most obvious features. The bat wing is essentially a modified hand — hence the Greek name Chiroptera, meaning "hand wing". The digits, except the thumb, are greatly elongated to support the flight membranes with the aid of a lengthened forearm. The thumb is usually largely free of the membrane and has a claw, although sometimes this is small; the second digit in most Megachiroptera also has a claw but otherwise claws or nails are lacking on the wings of modern bats. Feet are usually relatively small, with five toes, each with a strong claw.

The flight membranes, which in many bats also join the legs and tail, are extensions of the body integument or skin. The membranes are muscular and tough but very flexible, with a high concentration of blood vessels. Apparently hairless, the membranes do, in fact, have many short

► There are nearly 1,000 species of bats but only three of these feed on blood: vampire bats have given the rest of the bats an undeservedly bad reputation. The common vampire in this photograph is hopping—vampires land on the ground near their intended victims and walk and hop towards their feet.

Stephen Dalton/NHPA

transparent or translucent hairs, sometimes in bands or fringes, and the body fur may extend onto the inner part of the membrane.

Bats are the only mammals capable of true flight, an ability among modern vertebrates shared only with birds; other so-called "flying" mammals in fact glide with the aid of an outspread flexible membrane along the sides of the body and sometimes between the hindlegs and tail. Grounded bats scurry awkwardly but sometimes rapidly, although the common vampire *Desmodus rotundus* can walk, hop, and run with considerable agility using thumbs, wrists, elbows, and feet, while some free-tailed bats in the family Molossidae and the New Zealand short-tailed bat *Mystacina tuberculata* crawl and climb quite readily.

The wings are pulled downward by muscles on the chest and under the upper arm, and raised by other muscles and muscle groups on the back that act on the upper humerus and scapula (shoulder blade). This is very different from birds, where far fewer but relatively larger muscles on the chest provide the power for upward and downward movements of the wings. Further muscles along the arms in bats control the extension and

retraction of the wing, and its orientation in flight. Wing shapes vary quite widely. In general, bats that fly slowly through vegetation or other obstacles with very maneuverable flight have relatively short, broad wings, while fast-flying species that hunt in open spaces have longer, narrower wings.

Bats' ears vary widely in size and shape, some bats having exceptionally long ears. Many also have a tragus — a small projecting flap just inside the ear conch that obscures the ear opening. Contrary to popular belief, all bats have functional eyes: the eyes of fruit-eating species are large and adapted to poor light, but in most other species the eyes are relatively small and in a few they are nearly hidden in the fur. Several bat families have a noseleaf consisting of fleshy structures of the skin surrounding and surmounting the nostrils. Some bats have no tail or only the rudiment of a tail; in others the tail extends across and within the tail membrane or protrudes from its upper surface; and in some species the tail extends freely beyond the edge of the membrane. Many bats are drably colored in shades of brown or gray but some have striking color patterns.

▲ Bats fall into two very distinct groups: the herbivorous Megachiroptera and the largely insectivorous Microchiroptera, which make up about 80 percent of bat species. The long-eared bat is a typical microchiropteran.

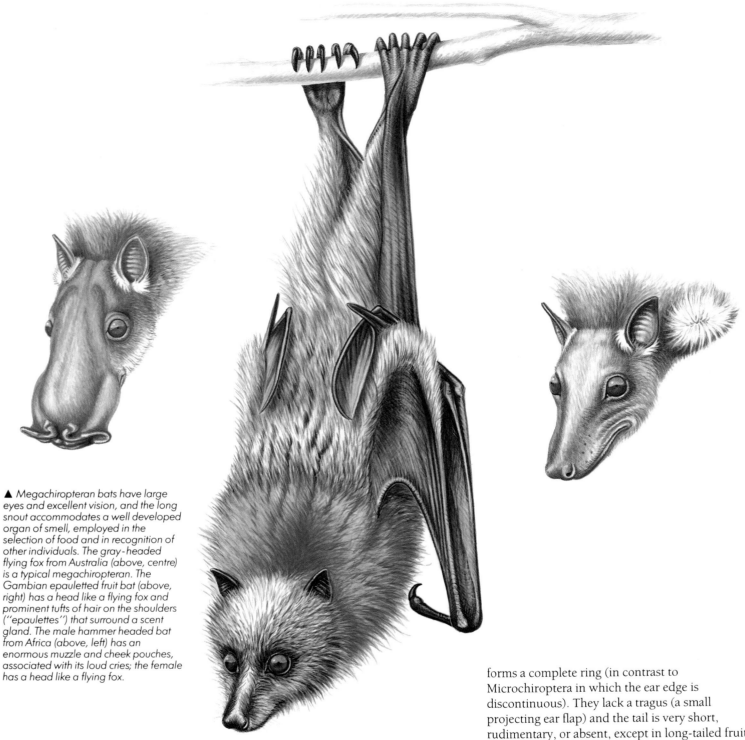

▲ Megachiropteran bats have large eyes and excellent vision, and the long snout accommodates a well developed organ of smell, employed in the selection of food and in recognition of other individuals. The gray-headed flying fox from Australia (above, centre) is a typical megachiropteran. The Gambian epauletted fruit bat (above, right) has a head like a flying fox and prominent tufts of hair on the shoulders ("epaulettes") that surround a scent gland. The male hammer headed bat from Africa (above, left) has an enormous muzzle and cheek pouches, associated with its loud cries; the female has a head like a flying fox.

OLD WORLD FRUIT BATS (MEGACHIROPTERA)

Old World fruit bats, flying foxes and dog-faced fruit bats belong to the one family Pteropodidae in which there are over 160 species. Found in the tropics and subtropics of the Old World (Europe, Asia, Australia, Africa), these are medium to very large bats, with a head and body length of 50 to 400 millimeters (2 to 16 inches) and a forearm of 40 to 230 millimeters (1½ to 9 inches). They have no noseleaf and a simple ear, the edge of which

forms a complete ring (in contrast to Microchiroptera in which the ear edge is discontinuous). They lack a tragus (a small projecting ear flap) and the tail is very short, rudimentary, or absent, except in long-tailed fruit bats, genus *Notopteris*, which have a 60 millimeter (2⅓ inch) tail outside the narrow tail membrane.

All species in this family live on fruit, flowers, and flower products. Only the genus *Rousettus* is known for certain to utilize a simple form of echolocation. Many species are brownish or blackish, sometimes with a brighter mantle or with a gray or silvery tinge; others have speckled ears and membranes or a facial pattern of white spots or stripes, perhaps to aid concealment in foliage. Most roost in trees or in dimly lit areas of caves. The fruit-eating species usually have large, flat-crowned grinding teeth and relatively short, strong jaws; those that eat nectar and pollen have long muzzles with lightly built jaws and teeth, and a long, extensible tongue with brush-like papillae.

INSECT-EATING BATS (MICROCHIROPTERA)

Mouse-tailed bats

Mouse-tailed bats belong to the family Rhinopomatidae, in which there are three species in one genus. Found from North Africa to southern Asia, these small to medium-sized bats have a length of 53 to 90 millimeters (2 to 3½ inches) and a forearm of 45 to 75 millimeters (1¾ to 3 inches). They have no noseleaf, but the muzzle is swollen with a transverse ridge above slit-like nostrils. The ears are large, joined at the base, and have a small tragus. Mouse-tailed bats have a long, thread-like tail, which is similar in length to the forearm and projects from the edge of the narrow tail membrane. Grayish brown to dark brown in color, these bats are insectivorous and adapted for life in arid and semi-arid regions. They are gregarious and colonial, often roosting in artificial structures.

Sheath-tailed bats

There are 49 species of sheath-tailed, sac-winged, pouched, tomb, and ghost bats in the family Emballonuridae. Widely distributed in the tropics and subtropics they are small to medium in size: head and body 37 to 135 millimeters (1½ to 5¼ inches) in length, with a forearm of 32 to 95 millimeters (1¼ to 3¾ inches). They have no noseleaf. The ears vary among species but are often joined at the base, with a small or moderate tragus. The tail is partially enclosed in the tail membrane with its tip protruding from the upper surface of the membrane. In some species there is a glandular sac in the wing membrane anterior to the forearm near the elbow or a glandular pouch on the throat. Most species are brown or grayish brown in color; some species are almost black, and ghost bats, genus *Diclidurus*, are white or grayish white. All species are insectivorous and roosts vary from caves to hollow trees and foliage.

Hog-nosed bat

The single species of hog-nosed or bumblebee bat belongs to its own family, Craseonycteridae, and is found in southwest Thailand. It is a very small species, with a head and body length of only 29 to 33 millimeters (1 to 1⅓ inches) and a forearm of 23 to 26 millimeters (1 inch). The hog-nosed bat has a muzzle with a low transverse ridge above the nostrils but no noseleaf. Its ears are very large with a swollen tragus. The tail membrane is extensive, but there is no external tail. Brown to reddish gray with a paler underside, it is insectivorous and roosts in caves. The sole species, *Craseonycteris thonglongyai*, was discovered as recently as 1973.

Merlin D. Tuttle/Bat Conservation International

It is the smallest known bat and is among the smallest mammals.

Slit-faced bats

There are 14 species in one genus of slit-faced, hollow-faced, or hispid bats (family Nycteridae), which are found in Africa, and Southwest and Southeast Asia. The total head and body length is 43 to 75 centimeters (1¾ to 3 inches) with a forearm of 35 to 66 millimeters (1⅓ to 2½ inches). The slit-faced bat has a complex noseleaf, with folds and outgrowths flanking a deep longitudinal groove. The ears are large and joined at the base with a small tragus. The tail membrane is extensive and encloses the tail, which has a small T-shaped cartilaginous tip at the edge of the membrane. In color, slit-faced bats are usually brown to reddish brown. Their diet includes large arthropods, such as spiders and scorpions, although the large slit-faced bat *Nycteris grandis* also takes small vertebrates. Slit-faced bats utilize a variety of roosts from caves to tree holes and even the abandoned burrows of other mammals.

False vampire bats

False vampire bats and yellow-winged bats belong to the family Megadermatidae, which is found in Central Africa, Southeast Asia and Australia, with five species in four genera. Medium to large in size, with a head and body length of 65 to 140 millimeters (2½ to 5½ inches) and a forearm of 50

▲ *In most microchiropterans the tail helps to support a membrane between the hindlegs, but the three species of mouse-tailed bats have no such membrane and the long, dangling tail gives them a peculiar appearance.*

Stephen Dalton/NHPA

▲ *The so-called false vampire bats of the Old World and the Australasian region have no resemblance to the true vampires of Central America. They are carnivores and, uniquely among the Microchiropteran bats, combine echolocation with good vision. They capture small mammals, such as mice, on the ground . . .*

to 115 millimeters (2 to 4½ inches), they have a conspicuous, long, erect noseleaf. The ears are large and joined at the base with a prominent bifurcated tragus. Although the tail membrane is extensive the tail is short or absent. Color varies from blue-gray to gray-brown; the flight membranes are pinkish white in the Australian false vampire *Macroderma gigas* and yellowish in the African yellow-winged bat *Lavia frons.* Their diet includes large insects and small vertebrates such as frogs, birds, rodents, and other bats, and they hang in wait for passing prey. They roost in caves, rock crevices, hollow trees, or foliage.

Horseshoe bats

There are 63 species of horseshoe bats in the family Rhinolophidae, found in the tropics, subtropics,

and temperate zones of the Old World. They are small to medium in size, with a head and body length of 35 to 110 millimeters (1⅓ to 3 inches) and a forearm of 30 to 75 millimeters (1¼ to 3 inches). They have a distinctive horseshoe-shaped noseleaf with a strap-like sella (flat structure) above the nostrils and the central part of the noseleaf, usually with an upright, triangular, cellular, and bluntly pointed posterior projection or lancet extending the noseleaf to the rear. The ears are relatively large but have no tragus. The tail membrane encloses the tail. Most species are brown or reddish brown. Generally tropical or subtropical, but with a few temperate species that hibernate in the winter, rhinolophids forage for insects near the ground or among foliage, roosting in caves, mines, hollow trees, or buildings.

Stephen Dalton/NHPA

Old World leaf-nosed bats

The family Hipposideridae contains 66 species of Old World leaf-nosed bats and occurs in the tropics and subtropics. There is a wide variation in size among species, ranging from 28 to 110 millimeters (1 to 4 inches) in head and body length. The noseleaf is similar to that of the horseshoe bats, but lacks the sella above and behind the nostrils and the upright lancet; it is usually rounded and not triangular at the rear. The noseleaf is sometimes complex. There is no tragus and the ears are moderate-sized. Most species are some shade of brown and in all species the tail membrane encloses the tail. Old World leaf-nosed bats eat a wide variety of insects, as their size variation might suggest, and generally roost in caves where they sometimes form large colonies.

Bulldog bats

Bulldog bats, sometimes called hare-lipped, mastiff, or fisherman bats, belong to the family Noctilionidae, in which there are two species in one genus. Ranging in length from 57 to 132 millimeters (2¼ to 5 inches) with a forearm of 54 to 92 millimeters (2 to 3½ inches), bulldog bats have a pointed muzzle with a pad at the end, and lips and cheeks that form pouches. They have no noseleaf and large, slender ears with a small tragus. The moderate tail membrane encloses the tail, with its tip emerging from the upper surface. The greater bulldog bat or fisherman bat *Noctilio leporinus* catches fish up to 10 centimeters (4 inches) in length (which it detects just beneath the water surface) by seizing them with its long, sharp claws and enormous feet. Bulldog bats roost in

▲ . . . *killing them with a bite to the head or neck. They also take large insects and frogs — some even catch other bats on the wing. Some species of false vampires have a wingspan of about 30 centimeters (1 foot).*

Merlin D. Tuttle/Bat Conservation International

▲ *Bulldog or fisherman bats use their long, rake-like feet to capture fishes and large insects at the surface of lakes or calm rivers. The short fur is water-repellent.*

▶ *The Californian leaf-nosed bat is a member of the New World family Phyllostomidae, which is characterized by an immense dietary range. This species eats insects and fruits.*

Jeff Foott/Bruce Coleman Ltd

caves, rock crevices, or hollow trees in the tropical regions of the New World where they are found.

Naked-backed bats

Naked-backed, moustached, or ghost-faced bats belong to the family Mormoopidae found in the southern United States and Antilles to Brazil, with eight species in two genera. Small to medium in size, they have a head and body length of 40 to 77 millimeters (1½ to 3 inches) and a forearm of 35 to 65 millimeters (1⅓ to 2½ inches). They have no noseleaf, but the chin and lips have complex plates and folds of skin. The large ears, sometimes joined at the base, have a complex tragus. There is an extensive tail membrane with about one-third of the tail protruding from the upper surface of the membrane. Brown or reddish brown in color, some wing membranes attach along the midline of the back, hence the name naked-backed. Insectivorous and often gregarious, naked-backed bats form moderate to large colonies in hot, humid caves. Most fly swiftly, often near the ground.

New World leaf-nosed bats

New World leaf-nosed bats in the family Phyllostomidae comprise 152 species in 51 genera. Bats of the New World tropics and subtropics, they vary in size from small to large: head and body measure 40 to 135 millimeters (1½ to 5 inches); forearm 25 to 110 centimeters (1 to 4 inches). Generally the muzzle has a simple, spear-shaped noseleaf, but the ears vary in size and shape, although all have a tragus. Some species have tails, and, when present, the tail is usually enclosed in the tail membrane with its tip sometimes projecting slightly beyond the edge of the membrane. Most species are reddish or reddish brown, some with white stripes on the face or back; one species — the white bat *Ectophylla alba* — is, as its name suggests, white. Few species are exclusively insectivorous. Many species are more or less omnivorous, eating insects, fruit, and in some cases they even prey on small vertebrates; others are frugivorous. Some species are nectar and pollen feeders, with long muzzles, reduced jaws and teeth, and an extensible tongue with brush-like papillae. The three species of vampire — *Desmodus rotundus, Diaemus youngi,* and *Diphylla ecaudata* — are specialized for an exclusive diet of blood. This family embraces almost the entire spectrum of bat food habits, except fish-eating, and its members exploit a wide variety of roosts from caves to trees and include the few genera (*Uroderma, Artibeus, Ectophylla*) that construct a rudimentary shelter.

Funnel-eared bats

The family Natalidae contains five species (in one genus) of funnel-eared or long-legged bats. Distributed from Mexico to Brazil and the Antilles, funnel-eared bats are small with a head and body length of only 35 to 55 millimeters (1⅓ to 2 inches) and a forearm 27 to 41 millimeters (1 to 1½ inches). Their nostrils are close together near the top lip and they have no noseleaf. The ears are large and funnel-shaped, with a variously distorted and thickened tragus. The extensive tail membrane encloses the tail and the hindlimbs are long and slender. Gray or yellowish to reddish brown in color, males have a curious bulbous "natalid" organ of unknown function on the forehead. All species are insectivorous and utilize caves, tunnels, or rock overhangs as roosts.

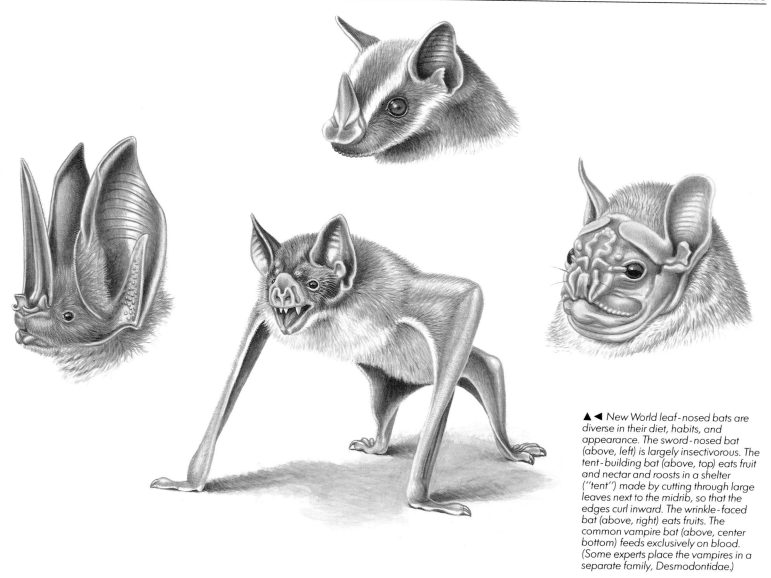

▲ ◄ New World leaf-nosed bats are diverse in their diet, habits, and appearance. The sword-nosed bat (above, left) is largely insectivorous. The tent-building bat (above, top) eats fruit and nectar and roosts in a shelter ("tent") made by cutting through large leaves next to the midrib, so that the edges curl inward. The wrinkle-faced bat (above, right) eats fruits. The common vampire bat (above, center bottom) feeds exclusively on blood. (Some experts place the vampires in a separate family, Desmodontidae.)

Smoky bats

Smoky bats or thumbless bats belong to the family Furipteridae found in Central America and northern and western South America. The two species in two genera are small, with a head and body length of 37 to 58 millimeters (1½ to 2¼ inches) and a forearm of 30 to 40 millimeters (1 to 1½ inches). They have large funnel-shaped ears with a small tragus, but no noseleaf. A rudimentary thumb is enclosed in the flight membrane, which extends to the base of a minute claw, and the tail is enclosed in a moderate tail membrane. Grayish or brownish gray in color, they are insectivorous and roost in caves.

Disc-winged bats

There are two species in one genus of disc-winged or New World sucker-footed bats in the family Thyropteridae, which is found from southern Mexico to Peru and Brazil. Both species are small: head and body 34 to 52 millimeters (1⅓ to 2 inches) long, and the forearm 270 to 380 millimeters (1 to 1½ inches) long. The long,

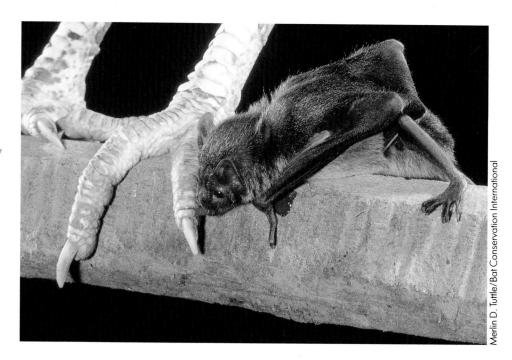

Merlin D. Tuttle/Bat Conservation International

▲ A vampire bat lapping blood from the foot of a fowl.

Merlin D. Tuttle/Bat Conservation International

▶ Disc-winged bats have an adhesive sucker on each thumb and foot and can hang from any one sucker. They nest in naturally curling leaves, such as those of banana trees.

slender muzzle has no noseleaf. The ears are large and funnel-shaped, with a prominent tragus. The extensive tail membrane encloses the tail, which projects slightly from the edge. There is a large sucker-shaped adhesive disc on a short stalk at the base of thumb and on the side of foot. In color they are reddish brown to light brown on the back, with a whitish or brownish underside. Disc-winged bats roost in family groups in a rolled leaf or frond in an unusual head-upward position; the adhesive discs on the thumbs and feet provide a grip, and the bats move to a new roost as the leaf unfurls.

Old World sucker-footed bat

The single species of Old World sucker-footed bat (family Myzopodidae) is found exclusively in Madagascar. It is a medium-sized bat with a head and body length of 57 millimeters (2¼ inches) and a forearm of 46 millimeters (1¾ inches). There is

no noseleaf. The ears are long and slender, with a small square tragus partially fused to the front edge of the ear and a curious mushroom-shaped process consisting of a short stalk supporting a flat expansion at the base of the back edge of each ear, which partially obscures the ear opening. The extensive tail membrane encloses the tail, which projects slightly beyond the edge. There is a sucker-shaped disc at the base of the thumb and the side of each foot. The toes are joined by webbing that extends almost to the base of the claws. Its habits are apparently similar to those of the disc-winged bats.

Vespertilionid bats

Vespertilionid bats (family Vespertilionidae), of which there are 350 species in 43 genera, are worldwide in distribution (except for polar regions and some oceanic islands). Species vary in size from very small to large, with head and body lengths of 32 to 105 millimeters (1¼ to 4 inches), and a forearm of 24 to 80 millimeters (1 to 3 inches). No species in this family has a noseleaf, but some have a transverse ridge. Ears vary in size and shape, with a short to long tragus. The tail membrane is usually extensive, enclosing the tail, which sometimes projects slightly from the edge. A few species have adhesive pads at the base of the thumb or foot. Most species are brown, grayish, or blackish, but some species are more brightly colored or have white spots or stripes. One species — the spotted bat *Euderma maculatum* — is patterned black and white, while another — the painted bat *Kerivoula picta* — has wing membranes patterned in black and orange. The members of this family occupy a wide range of habitats from semi-desert to tropical forest; species in temperate

▼ This photograph of a long-eared bat shows that the structure of a bat's wing is fundamentally the same as that of a human arm. There is a short upper arm, a longer forearm, and a tiny wrist. The thumb is a claw and the four long fingers support the wing membrane.

Press-Tige Pictures/Oxford Scientific Films

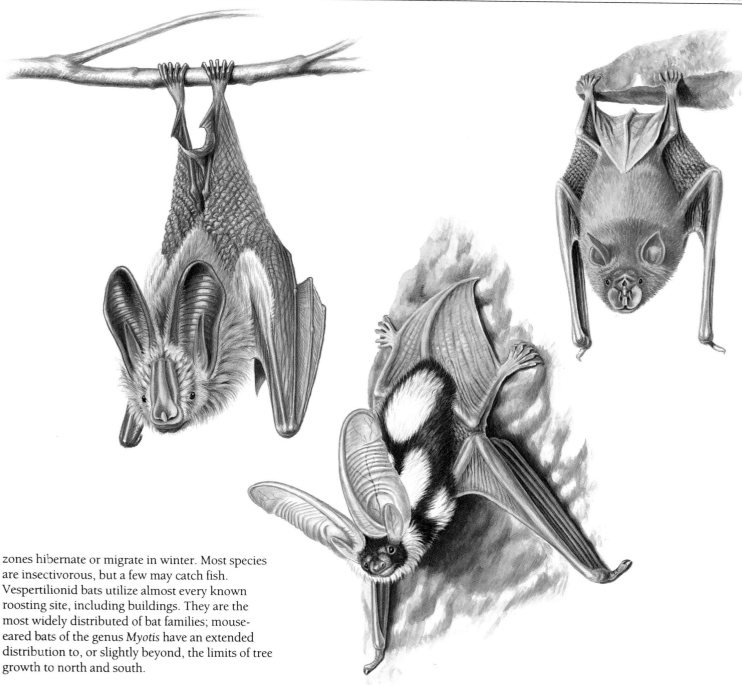

zones hibernate or migrate in winter. Most species are insectivorous, but a few may catch fish. Vespertilionid bats utilize almost every known roosting site, including buildings. They are the most widely distributed of bat families; mouse-eared bats of the genus *Myotis* have an extended distribution to, or slightly beyond, the limits of tree growth to north and south.

New Zealand short-tailed bats

As their name suggests, the two species of New Zealand short-tailed bat in the family Mystacinidae are found exclusively in New Zealand. (One species is thought to have become extinct recently.) They are small to medium in size with a head and body length of 60 millimeters (2⅓ inches) and a forearm of 40 to 48 millimeters (1½ to 1¾ inches). The front of the muzzle is obliquely truncate and there is no noseleaf. The ears are large and pointed, with a long narrow tragus. The wings are thick and leathery. The tail membrane is moderately long and thin, enclosing the short tail with its tip protruding from the upper surface of the membrane. The legs are short; the claws have a small additional talon; and the feet are wrinkled. These bats are grayish brown, brown, or blackish

brown in color; the fur is soft and velvety. Well adapted for moving about on the ground or in trees and chiefly insectivorous, New Zealand short-tailed bats possibly also eat fruit or flower products. They roost in caves, crevices, and hollow trees but apparently do not undergo prolonged hibernation. Conventionally thought to be related to vespertilionid bats and the free-tailed bats, some evidence suggests a closer relationship to the New World bulldog bats, naked-backed bats, and New World leaf-nosed bats.

Free-tailed bats

The family Molossidae comprises approximately 89 species in 13 genera and is found in both Old and New World tropics and subtropics, with some species of free-tailed bats extending into the

▲ Most bats have dull brown to gray fur but some species are strikingly colored, as in the yellow-winged false vampire bat (above, left), the vespertilionid spotted bat (above, center), and the orange horseshoe bat (above, right). Orange-colored individuals are not uncommon in otherwise drab-colored species of horseshoe bats.

Stephen Krasemann/Bruce Coleman Ltd

▲ *Many bats roost communally in caves, often in large numbers. It has been estimated that communities of the Mexican free-tailed bat, here seen emerging from a maternity cave, number more than 40 million. Some flying fox roosts shelter more than a million bats.*

temperate zones. There is great size diversity within this family, with some very small species and some large; head and body lengths vary from 40 to 130 millimeters (1½ to 5 inches), and forearms 27 to 85 millimeters (1 to 3⅓ inches). The muzzle is broad and, as in the New Zealand short-tailed bats, is obliquely truncate. The lips of many species are wrinkled, and there is no noseleaf. Ears are usually moderate in length, but thick and often joined at the base, with a small tragus. The flight membranes in some species are rather tough and leathery while the tail membrane is relatively narrow, enclosing the stout tail that projects considerably beyond its edge. The legs are short and strong. The body fur is usually brown or blackish brown, but wing membranes are sometimes whitish. One species — the hairless bat, genus *Cheiromeles* — is, as its name implies, effectively devoid of fur. The Mexican free-tailed bat *Tadarida brasiliensis* at least is migratory and a few other species enter short periods of winter inactivity in the more temperate parts of the range. Free-tailed bats are strong, fast flyers, catching insects on the wing. They roost in caverns, tunnels, hollow trees, foliage, under rocks or bark, or sometimes in buildings. Most species are highly gregarious and some form very large colonies.

A VARIED DIET

Approximately 70 percent of bat species feed on insects and other small arthropods such as spiders or scorpions; the size of their aerial prey ranges from gnats to large moths. Insects are captured in flight using the mouth; the tail membrane may be curled into a scoop from which the insect may be retrieved, or the insect may be deflected towards the mouth with the wing. Some bats glean insects from foliage, take them from the ground, or even skim them from the surface of water. Small insects can be consumed in flight, but larger items are carried to a nearby perch to be eaten. Large numbers of insects are consumed by bats, which thus play an important part in the control of insect populations. A large colony of bats may eat a substantial weight of insects annually. A few species of bats catch and eat frogs, lizards, small rodents, and birds, or even other bats, although this is not an exclusive habit, since insects are also taken. A small number catch fish, skimming over the water and seizing their prey with their long claws and strong feet.

The three species of vampire are exceptional in that they live on the blood of other animals. Vampire teeth are highly specialized, with the upper incisors and canines enlarged and razor sharp to inflict a small wound from which the blood is lapped. Vampire bats are found only in Central and South America, where local populations of the common vampire are thought to have increased greatly since European colonization and the introduction of cattle and horses.

Fruits, flowers, nectar, and pollen form the staple diet of Old World fruit bats or Megachiroptera. In the New World, where there are no megachiropterans, these food habits have evolved independently in many species of the Microchiroptera. The post-canine or grinding teeth of fruit-eating bats are usually low and flat-crowned (in contrast to the ridged and strongly cusped teeth of insectivorous species) and are sometimes reduced in number and size, especially in nectar-feeding bats. Fruit-eating and nectar-eating bats are confined to the tropics and subtropics, where food is available throughout the year. They rely upon a wide variety of fruits and flowers, and play a vital role as pollinators and seed dispersers for many plant species of importance commercially, or as food. Many such plants are adapted for bat pollination — their flowers open at night, are often white, creamy or greenish with a musky or sour scent, are shaped to facilitate landing and entry by bats, and hang free so that they can be easily reached, an example of coevolution between plants and bats. A few species of Old World fruit bats are known to eat leaves occasionally but the digestive system of bats is not adapted for herbivory, and this seems, therefore, to be a rare habit.

Jane Burton/Bruce Coleman Ltd

▲ Evolution is largely a matter of seizing opportunities. In the absence of megachiropterans from the New World, some microchiropterans have evolved into omnivorous or fruit and nectar eating species, such as the Barbados fruit bat shown here.

Merlin D. Tuttle/Bat Conservation International

◄ The spear-nosed long-tongued bat, a member of the Phyllostomidae, is able to hover like a hummingbird while it sips nectar with its long tongue. In much the same way as a bee, it fertilizes the flowers of some trees in the American tropics.

▼ *Gould's long-eared bat, an Australian species, gives birth to twins. When young, they may be carried by the mother. These two are suckling from the mother's functional teats; when she is flying, they reverse their position and bite onto a pair of false teats in the groin.*

Kathie Atkinson

REPRODUCTION AND DEVELOPMENT

Reproduction is generally seasonal, with birth and development coinciding with periods of maximum food abundance. In the tropics the young of some species are born shortly before the onset of the rainy season, although there may be more than one reproductive cycle annually; in the temperate zones birth occurs at the beginning of the summer months. Temperate-zone bats mate in the fall and during or at the end of hibernation. Viable sperm can be stored in the female or male reproductive tracts of bats throughout hibernation or winter; ovulation and fertilization occur early in the following spring. Some tropical species also store sperm but, in addition, delayed implantation or retarded development of the fertilized ovum can be used to ensure that sperm production and birth occur at favorable times.

Gestation periods in bats vary from 40 to 60 days in small species to as long as eight months in larger species. Generally only one young is produced, but twins occur regularly in some species and up to four or five young have been recorded for a few species. At birth the young are helpless but have strong claws and hooked milk teeth with which they cling to the mothers. At birth baby megachiropterans are relatively large and hairy, and have their eyes open, but most microchiropterans at birth are relatively small and naked, and their eyes are closed.

The females of many temperate and some tropical species congregate in nursery colonies to bear and raise the young. The site of such a colony may be traditional and used year after year. Usually the young are left behind while the mother forages and feeds, although in small species the young can begin to make short flights within three weeks. Nursing lasts from one to three months while the young learn to fly, hunt, and feed. Infant bats emit loud calls with patterns that permit their mothers to reunite and nurse only their own infants; at close range, odor is also a likely component of the reunion. In some species the mother and infant call back and forth to each other, and the infant increasingly matches its mother's frequencies and in this way possibly learns some aspects of adult echolocation calls. Female bats have one pair of thoracic teats but a pair of false teats occur in the groin region in a number of species; these may provide an additional hold for the young, which grow rapidly and attain sexual maturity late in their first year or in their second year.

HIBERNATION FOR SURVIVAL

Many bat species can regulate their body temperature (heterothermy), allowing it to fall while roosting during the day to reduce their energy consumption, while others, such as large tropical fruit bats, maintain a relatively constant body temperature (homeothermy). In the temperate zones torpidity is extended into hibernation during the winter months when food is scarce or unavailable; body temperature and metabolism are reduced to very low levels and survival depends on fat stored during the fall. Hibernation is not necessarily continuous and hibernating bats may wake and move to a different part of the hibernaculum or even to another site.

Michael Fogden/Bruce Coleman Ltd

A number of temperate species migrate seasonally to avoid the extremes of winter, returning in the spring. The European noctule *Nyctalus noctula* travels as much as 1,000 to 1,600 kilometers (620 to 1,000 miles) or, exceptionally, 2,000 kilometers (1,200 miles) when migrating. Some fruit bats also appear to follow the seasonal flowering and fruiting of their food plants. (Accidental dispersal also occurs and occasionally bats are found far from their usual range, having been blown for long distances by the winds or accidentally transported in ships or aircraft.)

ECOLOGY AND BEHAVIOR

Bats occupy a wide variety of roosts. Many bats are cave dwellers, but they are found also in mines, tunnels, culverts, tombs, ruins, under shallow rock overhangs, in cracks and crevices, under loose rubble, or in buildings. Others live under bark, in hollow trees, or hang in the foliage of trees, shrubs, or bushes. A few New World bats of the genera *Uroderma, Artibeus,* and *Ectophylla* make a primitive, tent-like shelter by biting the supporting ribs of palm fronds so that these collapse. Many bats are strongly gregarious, sometimes forming large colonies of many thousands of individuals in

a single cave or cave complex. Others live in small groups and a few appear to be solitary. Some form breeding groups or harems that may persist throughout the year, the males using visual,

Merlin D. Tuttle/Bat Conservation International

▲ The tiny Honduran tent bat is one of the few white bats. Like other tent-building bats, they cut away the connection between the edges of a palm frond and its midrib, causing the leaf to curl. Several bats roost in the "tent" that is created.

◄ An eastern pipistrelle in hibernation. The bat has become so cool that moisture has condensed on its fluffed-up fur.

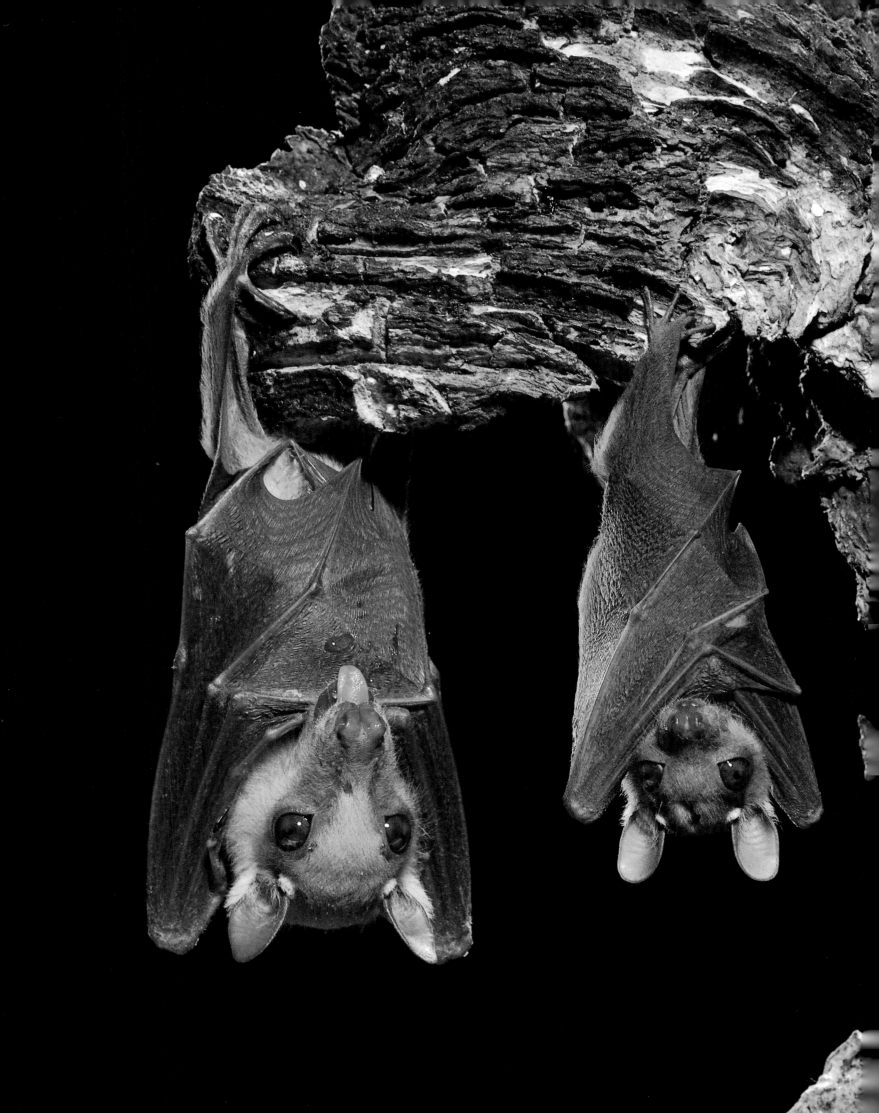

olfactory, or auditory signals to attract females or to discourage other males.

Bats appear to be relatively long-lived, with a maximum recorded age of 32 years in the little brown bat *Myotis lucifugus*. The European horseshoe bat *Rhinolophus ferrumequinum* has a known life-span of 26 years, and other species have been known to survive 10 to 20 years. The average life-span is, however, likely to be considerably less.

Bats are rarely a food item for any specific predator, but they can fall victim to a number of opportunistic hunters from large spiders to snakes, nocturnal birds, and mammals (even other bats).

CONSERVATION

Bats feature widely in mythology and folklore where, more often than not, they are quite unjustly associated with evil or, at best, the loathsome. Often there is an unjustified fear of bats, perhaps originating in their nocturnal and secretive habits or possibly from ignorance of these attractive and normally harmless animals. These unfounded fears and superstitions have adversely affected human attitudes towards bats and their conservation.

Occasionally fruit bats may cause local damage to crops but they are also important pollinating and seed dispersing agents and their destruction seems unjustified. The common vampire is a pest of domestic livestock in Central and South America, where it can also pose a health threat to humans. Insectivorous bats sometimes colonize roof spaces or other parts of buildings, but such colonies are usually harmless and, at worst, no more than a mild nuisance; often they are nursery colonies that will disperse at the end of the breeding season. If they have to be discouraged the only practical remedy is total exclusion by blocking off access points; extermination is useless since more bats move into the vacated area. In some countries, such as Great Britain, any unauthorized interference with bats is illegal. Bats can be

▲ *Mouse-eared bats in a nursery cave. However difficult it is to believe, each mother recognizes its own infant in such nurseries, apparently by a distinctive call and, perhaps, odor.*

encouraged by the provision of bat boxes on trees or in other suitable sites, or by other artificial roosts, and existing roosts can be protected: caves, for example, can be fitted with a carefully designed grille to prevent disturbance.

On the other hand, humans exploit bats. Large fruit bats are often cooked and eaten in parts of Africa, Asia, and the Pacific to the extent that the continued survival of some flying foxes of the genus *Pteropus* is seriously at risk. Some species are already extinct. In the past, bats have been a source of unusual remedies for various illnesses and, more recently, have been used in medical research. Bat manure has also been used for many years as a fertilizer. Perhaps the most bizarre involvement between humans and bats was a wartime project to use Mexican free-tailed bats *Tadarida brasiliensis* as self-propelled incendiary devices — the idea was hastily abandoned when, after some escaped, it became clear that they did not discriminate between friend and foe.

Bats are vulnerable to many threats, often the result of human activity, including habitat destruction, the loss of roosting sites, the indiscriminate use of pesticides, or simple persecution. Adverse public attitudes, sometimes reinforced by ignorant and uninformed comment in newspapers and magazines, only serve to propagate the myth that bats are harmful and dangerous. Bats are protected by law in many countries but legislation alone is not enough. There is a need for public education and a realization that bats are an outstanding example of evolutionary adaptation and an irreplaceable part of the global ecosystem. In Great Britain, such an approach, combined with legislation, has been remarkably successful in increasing public awareness of bats and of the need for their active conservation, and much progress is now being made in the United States.

ORIGINAL TEXT BY J. E. HILL;
REVISED BY GEORGE MCKAY

◄ *When a megachiropteran such as Wahlberg's epauletted fruit bat is roosting, it wraps the wings around its body and holds the head forward at a right angle to the chest. Microchiropterans, however, usually fold the wings at the side of the body and the head hangs down or is held at about a right angle to the bat's back.*

DO BATS CARRY RABIES?

Rabies virus or rabies-related viruses have been found in numerous species of New World bats; there are few reports from Europe, Asia, Australia, or Africa. Although many species can carry these viruses, their occurrence is confined chiefly to vampire bats (especially the common vampire) in Central and South America, where the sanguinivorous habits of these bats make them a potential risk. A very small number of human deaths over a long period in North America and still fewer elsewhere have ben attributed to bites but, since the vast majority of bats do not normally bite humans, the likelihood of infection is remote. Sick or moribund bats, however, should be handled cautiously.

FLYING IN THE DARK

As long ago as 1793 the Italian Lazzaro Spallanzani discovered that blinded bats were able to avoid obstacles when flying. A Genevan scientist, Louis Jurine, found soon after this that bats with obstructed ears were unable to do so. Both scientists concluded that hearing was important in nocturnal bat flight but could not explain why. Their work was disregarded and fell into oblivion, but in the early part of the twentieth century it was suggested that high frequency sounds were involved, and the development of sufficiently sensitive apparatus enabled an American scientist, Donald Griffin, to establish in 1938 that this was so. Almost coincidentally the Dutch zoologist Dijkgraaf arrived at a similar conclusion after studying the faint sounds sometimes heard when bats are flying, but it was not until the Second World War ended that his work became known. Most of the sounds are ultra-sonic and therefore beyond the normal range of human hearing of 20 hertz to 20 kilohertz; most echo-locating bats use frequencies of 20 to 80 kilohertz, but some range as high as 120 to 210 kilohertz.

The use of high frequency echolocation for navigation, hunting, and catching prey is confined to the Microchiroptera. The sounds are produced in the larynx and emitted through the mouth or through the nostrils in the case of bats with noseleaves, the leaf apparently serving to modify, focus, and direct the beam of sound. Echoes received by the ears provide information that can be processed by the brain to provide the bat with an interpretation of its surroundings and the location of flying or resting prey. The Megachiroptera lack echolocation of this type, but a few species of rousettes of the genus *Rousettus* produce orientation sounds that are partially audible (5 to 100 kilohertz) by clicking the tongue. Unlike other

▼ Pulses of ultrasound emitted by a bat spread outward from its head like ripples in a pond (below, right). The strength of the reflected vibrations gives information on the distance of the prey, while slight differences in the time taken for the reflections to reach each ear give information on its direction. Pulses of ultrasound are often complex. This sonogram of the relatively simple emission of a horseshoe bat (bottom) shows that the sequence begins with a series of constant tones, each of which glides rapidly to a lower frequency (through about a quarter of an octave). As the bat homes in on its prey, the pulses are emitted more rapidly, their frequency increases slightly, and the constant tone almost disappears.

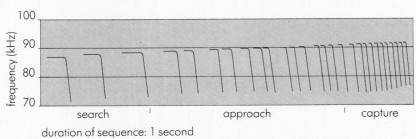

duration of sequence: 1 second

megachiropterans, these bats roost in the darker parts of caves and use this form of echolocation to navigate in and out of the caves, though they orientate visually in better light.

The ultrasonic pulses vary in duration from about 0.2 milliseconds to 100 milliseconds and may incorporate from one to five harmonics or overtones. By human standards of loudness the sounds may be very powerful and of high intensity. Frequency modulated (FM) signals sweep in a shallow curve or sharply through a range of frequencies, usually downwards; other signals are emitted at a constant frequency (CF). Most bats vary their acoustic repertoire, using both FM and CF patterns in varying degrees and combinations.

Stephen Dalton/NHPA

Some use CF signals of moderate to long duration to establish target movement and speed by means of frequency changes in the echo (Doppler effect). Typically, a hunting bat will search using an emission rate of 5 to 10 pulses a second. Once a flying insect has been detected the pulse rate increases to 15 to 50 a second, this approach phase being terminated by a stage in which the pulse rate may reach 200 pulses a second, this high rate providing the bat with continuous information about the target.

Some moths have a simple ear on the thorax that can detect bat echolocation sounds and, so warned, these moths are able to take evasive action. Other moths produce a train of ultrasonic clicks that mimic bat calls and so interfere with the bat's echolocation system. On the other hand, it is suspected that some long-eared bats can locate their prey by listening for its sounds; certainly the fringe-lipped bat *Trachops cirrhosus* is quite adept at discerning the social calls of the pond frogs that it hunts.

Megachiropteran bats also produce low frequency audible calls of many kinds. The epauletted fruit bat *Epomops franqueti* honks metallically during the night, and male hammer-headed fruit bats *Hypsignathus monstrosus* produce low, throaty, and intense metallic calls to attract females. Squeaks, squawks, and screams in the roost or around a fruiting tree may express annoyance or disputes over space.

▲ *To be able to chase and capture a moth in the dark, like this horseshoe bat, seems an astonishing feat. Even more remarkable is the fact that the bat simultaneously recognizes all the other objects in its vicinity—the ground, trees, bushes, rocks, other bats, and owls. It "sees" by means of ultrasound.*

PRIMATES

ORDER PRIMATES
• 11 families • 60 genera
• c. 200 species

SMALLEST & LARGEST

Red mouse lemur *Microcebus myoxinus*
Head–body length: 10.1 cm (4 in)
Tail length: 12.8 cm (5 in)
Weight: 35 g (2 oz)

Gorilla *Gorilla gorilla*
MALES
Height: 156 cm (61½ in)
Weight: 175 kg (385 lb)

CONSERVATION WATCH
!!! The 13 critically endangered species are: hairy-eared dwarf lemur *Allocebus trichotis*; golden bamboo lemur *Hapalemur aureus*; broad-nosed gentle lemur *Hapalemur simus*; Tattersall's sifaka *Propithecus tattersalli*; black-faced lion tamarin *Leontopithecus caissara*; golden-rumped lion tamarin *Leontopithecus chrysopygus*; golden lion tamarin *Leontopithecus rosalia*; golden-bellied capuchin *Cebus xanthosternos*; yellow-tailed woolly monkey *Oreonax flavicaudus*; Mentawai macaque *Macaca pagensis*; Tonkin snub-nosed monkey *Rhinopithecus avunculus*; white-rumped black lemur *Trachypithecus delacouri*; silvery gibbon *Hylobates moloch*.
!! 29 species are endangered, including: ruffed lemur *Varecia variegata*; indri *Indri indri*; aye-aye *Daubentonia madagascariensis*; buffy-headed marmoset *Callithrix flaviceps*; cotton-top tamarin *Saguinus oedipus*; Central American squirrel monkey *Saimiri oerstedii*; Preuss's monkey *Cercopithecus preussi*; Japanese macaque *Macaca fuscata*; Moor macaque *Macaca maura*; lion-tailed macaque *Macaca silenus*; drill *Mandrillus leucophaeus*; grizzled leaf monkey *Presbytis comata*; Douc langur *Pygathrix nemaeus*; black snub-nosed monkey *Rhinopithecus bieti*; pig-tailed snub-nosed monkey *Simias concolor*; black gibbon *Hylobates concolor*; gorilla *Gorilla gorilla*; pygmy chimpanzee *Pan paniscus*; chimpanzee *Pan troglodytes*.
! 54 species are listed as vulnerable, including proboscis monkey *Nasalis larvatus* and orang utan *Pongo pygmaeus*.

The order Primates contains around 200 species of mainly arboreal, keen-sighted, intelligent animals: lemurs, monkeys, apes, and their relatives. They are a diverse lot—just how diverse we have only come to realize in the past thirty years, when field studies of primates really began to blossom. Most unexpected of all, perhaps, has been that completely new species of primates have continued to be discovered right up to the present.

DIVERSE MAMMALS

Surprisingly, for one of the most intensively studied of all orders of mammals, the primates are difficult to define. Nearly all have hands and feet modified for grasping; flat nails, rather than claws, on at least some of their digits; ridged friction pads on the undersurfaces of the ends of their digits; and highly sensitive nerve-endings in these digit pads. Males have testes that are permanently descended into a scrotal sac, and a penis that hangs free from the abdominal wall. Primates have long gestation periods, slow growth, long infant and juvenile dependency periods (during which there are opportunities for learning), and long life-spans. They have big brains. Most of these features are not unique to primates—arboreal marsupials, especially the opossums of the Americas, have the same adaptations of the hands and feet; fruit bats have a similarly complex visual system; carnivores, dolphins, and elephants have big brains—but taken together, these features do define primates.

Most primates are tropical animals—specifically, they live in tropical rainforests. But some species live in other forest types (sclerophyll forests, mangroves, even coniferous forests); some do not live in forests at all (baboons and other monkeys from the savannas and woodlands of Africa; langurs from the dry thorn-scrub of India); and some live in temperate regions where there is snow for part of the year (macaques in Japan, northern China, and the Atlas Mountains; snub-nosed monkeys from the mountains of southwestern China).

Specialists now generally classify the order primates into two suborders: Strepsirrhini (lemurs and their relatives, with five or six families) and Haplorrhini (tarsiers, monkeys, apes, and humans, with six to twelve families). The best classification is not by any means agreed on, as the inter-relationships of some groups, such as the New World monkeys and the lemurs, are still uncertain.

Members of the Strepsirrhini, lemurs can easily be recognized among primates by their moist, dog-like snout (the rhinarium). Most of them have long, pointed snouts. Almost all strepsirrhines possess two remarkable features: a "toilet claw", a long claw on the second toe of the foot, and a "dental comb", a row of incisor and canine teeth, six teeth in all, across the front of the lower jaw, compressed and pointing forward. Both specializations are used for grooming; whether they have any other functions in some species is unclear, but the dental comb is, in some species, also used for scraping resin, part of their diet, off the bark of trees.

The Haplorrhini, which include monkeys, apes, and humans, have a dry, sparsely haired nose instead of a rhinarium. Their eyes are also different. Most of them (not all) are non-seasonal breeders—the females have sexual cycles all year round. Their placenta is of a type that has a very close vascular connection between maternal and embryonic tissue, quite different from the placenta of strepsirrhines. Each month, the wall of the uterus develops a special tissue called endometrium in preparation for implantation of an embryo, and if no fertilization takes place the endometrium is shed. Some haplorrhines—apes, humans, and many Old World monkeys—have very extensive endometrium development, and its monthly shedding is accompanied by loss of blood: menstruation. No other mammals menstruate.

Humans, apes, and Old World monkeys are classified as Catarrhini, which is a major division of the Haplorrhini. Members of this suborder have two, not three, premolars in each half of each jaw; nostrils that are close together and directed downwards; and opposable thumbs. Many other details of anatomy, physiology, and biochemistry unite the Catarrhini and distinguish them from the other suborder of the Haplorrhini, the Platyrrhini or New World monkeys.

Living catarrhines divide easily into two superfamilies: one contains the Old World monkeys (Cercopithecoidea), and the other the apes and humans (Hominoidea). Hominoids have broad chests, short lumbar spines, and no tails; cercopithecoids have a narrow chest, a long lumbar region, and at least a short tail.

LEMURS AND ALLIES

As "lower primates," lemurs were long neglected by scholars in favor of monkeys and apes. For years the French primatologist Jean-Jacques Petter was working almost alone on the island of Madagascar (now the Malagasy Republic) studying the behavior and ecology of its lemurs, but, as their amazing diversity has become clear, lemurs have at last become fashionable. French, American, British, German, and, above all, Malagasy scientists have begun intensive work on lemurs. Many lemurs are increasingly threatened by habitat destruction for cultivation and cattle grazing; some are very localized in their distribution, and two species are so localized they were only discovered in 1987 and 1988, when they were already on the verge of extinction. Five families of lemurs are found on Madagascar; the sixth family of the Strepsirrhini is Loridae, the lorises and bushbabies.

Dwarf and mouse lemurs

Dwarf and mouse lemurs, known as cheirogaleids, are, in the main, solitary; females occupy overlapping home ranges, and males defend territories that overlap with females' ranges. The smallest forms, such as the mouse lemurs *Microcebus*, the smallest of all primates at about 30 to 60 grams (1 to 2 ounces), are mainly insectivorous; the larger ones, such as the dwarf lemurs *Cheirogaleus*, eat fruit as well as insects.

"True" lemurs

The ring-tailed lemur *Lemur catta* is about a meter in length—more than half of which is the tail—and weighs 2.5 to 3.5 kilograms (5½ to 7½ pounds). Ring-tailed lemurs live mainly on the floors of open forests in the dry country of southwestern Madagascar. They usually form troops of about 12 to 20, in which females are dominant to males—females can displace males in disputes, or from favorite foods or sitting-spots. They are territorial: the troop occupies an area that can be as little as 5 or 6 hectares (12 or 15 acres), which it defends against other troops. The males in a troop have glands on the wrist provided with horny spurs, which they rub onto small trees with an audible "click", leaving a slash in the bark impregnated with scent. These spur-marks inform other lemurs, of the same and of neighboring troops, which male has been there and which troop "owns" that bit of land. Like all lemurs, ring-tailed lemurs are seasonal breeders; most mating occurs within a two-week period during April, but females who do not conceive then may ovulate again a month later. The gestation period is 136 days after which a single young is born. Young ride around clinging to their mother's fur at first, then gradually become independent, and are sexually mature at 19 to 20 months.

Until 1988 the five species of the genus *Eulemur* were included in the genus *Lemur*, but they differ in many ways, such as lacking the arm glands. They

are much the same size as the ring-tailed lemur, but tend to be brown or black, often with prominent cheek whiskers, ear tufts, and facial markings. They are distributed throughout Madagascar in both rainforest and dry forest; they tend to be more arboreal than *Lemur* and to live in smaller troops. Other members of the family are the gentle lemurs (*Hapalemur*) and the ruffed lemurs (*Varecia*).

Sportive lemurs

Sportive lemurs of the genus *Lepilemur* live throughout Madagascar in all habitats from rainforest to desert, where they cling to the stems

▲ In the early morning, ring-tailed lemurs go to the tree-tops and sit facing the rising sun, arms spread, as if sun-worshipping. In fact, they are warming up after the cold night. In the heat of the day, as here, they shelter in the shade of the middle canopy. During the rest of the day they are mainly seen on the ground.

Walt Anderson/Tom Stack & Associates

Indrids

The black and white indri *Indri indri,* at over 6 kilograms (13 pounds) the largest surviving lemur, is also the only one that is virtually tailless. It has long hindlegs, and moves by jumping from tree to tree. A pair of indris live with their offspring in a large range of 18 hectares (44 acres); they share part of it with neighboring pairs, but keep most of it exclusive as a territory. Indris are entirely diurnal. Their loud wailing cries float through the forest and enable neighboring groups to avoid each other; they also rub their cheeks, which are presumably glandular, on branches. The other indrids, the diurnal *Propithecus* (the sifakas) and nocturnal *Avahi,* are smaller and have long tails.

The aye-aye

The aye-aye *Daubentonia madagascariensis* is the only strepsirrhine that lacks the dental comb and toilet claw. The incisors, reduced to a single pair in each jaw, are huge and open-rooted, so that they grow throughout life; they are used for gnawing away the bark of trees to get at the grubs on which the aye-ayes feed. Only the hallux (great toe) has a flat nail; all other digits have long curved claws. The middle finger is wire-thin and is inserted into crevices in trees to extract or pulp grubs. Aye-ayes weigh about 2.8 kilograms (6 pounds) and are about 800 millimeters (30 inches) long, half of which is the bushy tail; they are black, with coarse shaggy fur and huge ears and eyes. They are nocturnal, solitary, and secretive.

Lorises and bushbabies

These lemur-like creatures, the Loridae, are found in Africa south of the Sahara and in Sri Lanka, South India, and Southeast Asia, but they differ little from the Malagasy lemurs (the aye-aye excepted).

The Loridae are divided into two subfamilies, Lorinae and Galagoninae. They are very distinct and some authorities classify them as separate families. The Lorinae are short-limbed and short-tailed; they move quadrupedally in a slow, gliding fashion, using their strong hands and feet to grip branches. They are the lorises of South and Southeast Asia, and the potto and angwantibo of Africa. In 1996 a new genus and species, *Pseudopotto martini,* was described in Cameroun. The other subfamily is the Galagoninae, the bushbabies or galagos of Africa; they have long bushy tails and long hindlegs, and move by jumping. Whereas Lorinae are entirely forest-living, Galagoninae live in both forest and bush country, from Somalia and Senegal to the Cape of Good Hope. Most lorids seem to live solitary lives, though sometimes groups of females nest together and share a home range; the males occupy territories that overlap the range of females. The larger species, such as the potto and the bigger bushbabies, eat mainly fruit; the smaller ones, mainly insects. The needle-clawed bushbaby *Euoticus elegantulus* of Cameroun and Gabon, eats the resin of forest trees.

▲ *The indri (top) is a very rare lemur. It is now protected in the mountain rainforest reserve of Perinet in northeastern Madagascar. Ruffed lemurs are the largest of the Lemuridae and live in the eastern rainforests of Madagascar; the red ruffed subspecies (above, bottom) is confined to the Masoala Peninsula at the northern end of this range.*

of spiny succulents. The largest species has a head and body length of about 300 millimeters (12 inches), and weighs 900 grams (2 pounds); the smallest is 250 millimeters (10 inches) long and weighs only 550 grams (1 pound 3 ounces); their tails are longer than the head and body in some species, shorter in others. They live solitary, nocturnal lives, feeding on leaves and flowers in small territories that are vigorously defended against their neighbors of the same sex, but the territories of breeding males and females overlap.

TARSIERS

One genus of the Haplorrhini that does not look much like a monkey at all is the tarsier (genus *Tarsius*). If anything, it looks rather like a rat-tailed bushbaby, with its long hindlegs and long tarsal region, long skinny fingers and toes, huge eyes, and big ears (which explains why in the past it was classified alongside bushbabies and other lemurs rather than with monkeys and apes, where it belongs). Tarsiers live in island Southeast Asia. There are several species: one on Sumatra, Bangka, Belitung, the Natuna Islands, and Borneo; one on Leyte, Samar, Bohol, Dinagat, and Mindanao in the Philippines; and at least three on the central Indonesian island of Sulawesi, the Sangihe Islands, and Peleng. Tarsiers are nocturnal; their eyes are truly enormous, and in the most specialized

species (*T. bancanus*, from the western Indonesian Islands) the orbits are so flared that the skull is actually broader than it is long.

Tarsiers live in pairs; they feed on insects, lizards, and other small vertebrates. They are inconspicuous, though not uncommon, especially in secondary forest. Their Indonesian name speaks volumes about their way of life and their general persona: *binatang hantu*—ghost animal.

Tarsiers form a subgroup of Haplorrhini all by themselves. To the other subgroup, the Simiiformes, belong all the others, sometimes called "anthropoids": monkeys, apes, and humans. Their big brains and great intelligence distinguish them from other primates. The first division is between the New World monkeys (Platyrrhini) and the rest.

▲ *The nocturnal gray mouse lemur (top, left) is still common in the dry forests of western Madagascar. The aye-aye (top, center) is an enigmatic rainforest lemur; even its exact range is poorly known, but it is certainly very rare. The Philippine tarsier (top, right), a small, nocturnal, carnivorous primate from Southeast Asia, looks very much like a lemur but is in fact a primitive haplorrhine. The angwantibo or golden potto (bottom) is a small lorid of West Central Africa, where it is uncommon. It lives on insects and their larvae, including noxious caterpillars. It weighs up to 300 grams (12 ounces).*

A CONSERVATION CONCERN

Madagascar is ecologically diverse: there is a wide central grassy plateau, and down its eastern flank runs a belt of rainforest, whereas the western side of the island is covered with dry sclerophyll forest. In the south is a semi-desert region, where spiny succulents are the main vegetation; there is a small pocket of rainforest in the northwest; and in the far north is the forested Amber Mountain. Each of these zones has its own lemurs: most of the widespread genera have at least a rainforest species and a dry-forest species, and sometimes arid-zone, northwestern, and Amber Mountain species as well. But very few lemurs are at all common today. Mouse lemurs and other small forms survive well in small, remnant forest patches; ring-tailed lemurs and white sifakas are well-protected in a few strict nature reserves; black lemurs are revered as sacred in some villages in their range in the northwestern rainforest zone. The indri has a small range, but part of it is a nature reserve visited by tourists; *Lepilemur, Phaner, Cheirogaleus,* and some species of *Eulemur* and *Hapalemur* are not endangered, though they are becoming more and more restricted as their forests are cut down.

But the conservation status of some lemurs is desperate. Twenty years ago about half a dozen aye-ayes were caught and released on a small island to try to ensure their survival, as they are irrationally feared and so deliberately killed by some village communities; it is now known that the species still survives in the eastern rainforest zone on the mainland, but very sparsely and in small numbers. The hairy-eared dwarf lemur *Allocebus trichotis* was long known from only three specimens, collected in the 1870s, and was feared extinct, until a fourth, kept as a pet, was found in 1965. Only in 1989 were specimens seen in the wild. The broad-nosed lemur *Hapalemur simus* was also feared to have been extinct since 1900, but in 1972 surviving populations, of not many more than 200 or 300 individuals, were found in the southeast; in 1985 it was discovered that about half of these were not of that species at all, but a new species, now known as the golden bamboo lemur *Hapalemur aureus.* Nearly as rare is the newly discovered Tattersall's sifaka *Propithecus tattersalli,* which has a few scattered populations of perhaps a couple of hundred individuals in the far northeast.

The Malagasy government is seeking urgent international aid to help it in ambitious education, village relocation, nature reserve, and conservation-for-development programs; the will is there, and the expertise is available, but the finance is critically lacking—and the exploding human population of Madagascar threatens to overwhelm even the best-protected areas.

▶ *The white sifaka lives in the dry forests of western Madagascar in small family groups that generally consist of more males than females.*

▼ *The rare and mysterious aye-aye eats fruits by gnawing through the outer husk or rind and scooping out the pulp with rapid movements of its thin, wiry, long middle finger.*

Frans Lanting / Minden Pictures

▲ The red uakari (top, left), the only New World monkey without a long tail, is a surprisingly agile leaper. The red howler ((top, right), the largest of the howler monkeys, lives in small territorial groups high in the trees. The cotton-top tamarin (bottom left) is a now-rare species of marmoset-like primate from Panama. Douroucoulis or night monkeys (bottom center) belong to the only genus of nocturnal monkeys. The muriqui or woolly spider monkey (bottom right) is the largest and rarest New World monkey. Only about 300 remain in the dry forest of southeastern Brazil.

NEW WORLD MONKEYS

Platyrrhines, or New World monkeys, are found in South and Central America, almost entirely in rainforest. Externally, they can be distinguished at once by the nose: the nasal septum is broad, so that the nostrils point sideways. Also, their thumbs are not markedly opposable to their other fingers, and some of them grip objects as readily between forefinger and third finger as between thumb and forefinger. The larger species have prehensile tails; no other primate has a prehensile tail.

The problem of how these monkeys got to the

Americas is one that has puzzled paleontologists, because their remains first turn up in early Oligocene deposits of about 35 million years ago, when South America was already separated from Africa, where their closest contemporary relatives have been found.

Marmosets and tamarins

Marmosets and tamarins range from the tiny pygmy marmoset *Callithrix pygmaea,* at 125 grams (4½ ounces) the world's smallest living monkey, to the golden lion tamarin, which weighs 600 grams (21 ounces). They all have claws on all digits except the great toe (which bears a nail), making them unique among monkeys. Another unusual characteristic is that, except for *Callimico,* all marmosets tend to have twin births. The larger species eat mainly fruit, the small ones insects; many eat gums and nectar.

There is a curious puzzle about their typical social organization: in captivity, all species seem to be most easily kept in pairs, and the young will not breed until they are removed from their parents cage. In the wild, pairs have been seen in some species, but tamarins seem more usually to live in quite large groups, of a dozen or more. What is more, the sex ratio in larger groups is often quite skewed, with more males than females.

The best analogies to marmosets' social systems come not from other primates, but from some birds, like Australian white-winged choughs and Florida scrub-jays, which exhibit "helper systems". The young stay with their parents well beyond maturity, and help in rearing the next clutch of offspring, and the next. What finally persuades them to leave is unclear, but in some birds the females tend to emigrate earlier than the males.

As nearly all marmosets bear twins, the potential load on the mother is enormous. When pairs are kept together in captivity, the male is the one who does most of the carrying of the infants, playing with them, and generally "babysitting"; in general, the female takes the young mainly for suckling. When the previous year's young are left in the group it is they, particularly the males, who perform more of the babysitting activities.

Night monkeys and titis

Whether night monkeys, of the genus *Aotus,* and titis, of the genus *Callicebus,* the other pair-living platyrrhines, also have "helper systems" is not known, but they do not have twins. A pair of titis will sit side by side, tails intertwined; the night monkeys seem less closely bonded. It used to be thought that there was only one species of night monkey, *Aotus trivigatus,* but not only are there obvious color differences from one area to another, there are also striking differences in chromosome: those from southern Colombia have 46 chromosomes, while those from the north of the same country have 54, 55, or 56. Five to ten species of

Sullivan & Rogers/Bruce Coleman Ltd

▲ *A group of red howler monkeys howl to defend their ranging area in forested areas of the Llanos of Venezuela.*

Aotus are now recognized. Evidence has been accumulating that *Callicebus* also has many more species than was once thought.

Other New World monkeys

Squirrel monkeys, of the genus *Saimiri,* are small greenish monkeys with white faces and black muzzles; females weigh 500 to 750 grams (18 to 26 ounces); males weigh a kilogram or more. These monkeys are unusual in that they are seasonal breeders. They live in large troops of 20 to 50 or more, with many females and only a few males. In the breeding season, lasting three to four months, the males put on fat and become very aggressive as they compete for mating opportunities.

The prehensile-tailed howler monkeys, of the genus *Alouatta,* are widespread from southern Mexico to northern Argentina. The "howls" are troop spacing calls, which have been known to carry as much as 5 kilometers (3 miles)! Different species' calls sound different: zoologists Thorington, Rudran, and Mack, after studying the small black species *Alouatta palliata* in Panama, and the large reddish *A. seniculus* in Venezuela, said: "The howls of a distant troop of *A. palliata* sound rather like the cheers of a crowd in a football stadium, whereas the howls of *A. seniculus* are much like the sound of surf on a distant shore or a roaring distant wind." Most species of howlers live in troops of 10 to 30 consisting of several members of both sexes.

Spider monkeys, of the genus *Ateles,* are much more active animals than howlers; their troops are also much larger and split up into small foraging parties of unstable composition from day to day.

Other New World monkeys are the strangely adorned sakis and uakaris, the intelligent capuchins and Humbold's woolly monkeys, and the endangered yellow-tailed woolly monkey.

▼ The range of the diana monkey or diana guenon (below, top) is now confined to the diminishing high forests of West Africa. Male mandrills (bottom, left) are brightly colored; in females and the young, these colors are more muted. Mandrills live in Gabon and neighboring countries in Central Africa. The Celebes black "ape" or Celebes macaque (center, right) is one of about seven species of macaques found on the island of Sulawesi, central Indonesia. In the gelada (bottom, right), only the adult male has the long mane, but both sexes have a bright red patch of bare skin on the chest.

OLD WORLD MONKEYS

The two major groups of Old World monkeys are usually referred to as subfamilies of a single family, Cercopithecidae, but many authorities now prefer to recognize two full families: the Cercopithecidae and Colobidae. Colobid monkeys tend to be specialized leaf-eaters. Like ruminants or kangaroos, they have stomachs with fermentation chambers containing bacteria that break down cellulose into short-chain fatty acids and so make extra nutrients available from the monkey's food. Cercopithecids have simple stomachs, and deep food-storage pouches in the lining of the cheeks.

Guenons

Cercopithecus monkeys, often called guenons, are a marvellously varied and brightly colored lot, divided into a number of species-groups. In rainforest regions members of three species-groups occur together, living high in the trees. In Gabon the local representatives of the three groups are the greater spot-nosed monkey *C. nictitans* (males weigh 6.6 kilograms, females 4 kilograms, 15 and 9 pounds); the much smaller moustached monkey *C. cephus* (males weigh 4 kilograms, females under 3 kilograms, 9 and 6½ pounds); and the intermediate-sized crested guenon *C. pogonias* (males weigh 4.5 kilograms, females a little more than 3 kilograms, 10 to 6½ pounds).

Each species lives in a small troop consisting of one adult male, three or four adult females, and their offspring. The troops are territorial, but normally a troop of monkeys consists of not just one of the species, but two or even all three. The troops of the three species have their usual social structure, but have more or less merged their separate identities together so that they move around, sleep, and feed together—perhaps for years. In some cases, the males of the two smaller species do not even bother to make their characteristic spacing calls, but rely on the deep booming call of the big spot-nose male to maintain the joint territorial space. Jean-Pierre and Annie Gautier have studied this remarkable symbiosis intensively in forests in northeastern Gabon. In one troop they studied, the male crested guenon was the leader: he made the spacing calls, and gave the warnings against birds of prey, whereas the male moustached monkey uttered the warning calls for terrestrial predators.

Baboons and their relatives

Other species of African monkeys have quite different social structures. The gelada *Theropithecus gelada,* a large baboon-like monkey from the grasslands of the Ethiopian plateau, is very sexually dimorphic males weigh over 20 kilograms (44 pounds), females about 13 kilograms (29 pounds), and males have enormous manes. A troop consists of a male and one or a few females ("one-male groups' or "harems"), like the guenons, but in this species the troops associate together to form huge herds, sometimes of several hundred, which forage together on the cliff-tops.

The gelada is sometimes considered a kind of baboon. Other baboons include the hamadryas baboon *Papio hamadryas,* which like the gelada lives in one-male groups, and the savanna baboons *Papio cynocephalus* and relatives, which live in troops ranging from about ten to several hundred, with many adult males as well as females. Hamadryas and savanna baboons hybridize where their ranges meet, in northern Ethiopia, and there is some dispute whether they are better classified as separate species or not.

In savanna baboons, both males and females are organized into dominance hierarchies. Dominant animals get priority of access to scarce resources, such as favored foods, and they lead troop movements, sometimes taking the major role in the troop's protection; dominant males do most of the mating, and dominant females, too, appear to be more reproductively successful, as their offspring have a higher survival rate.

Among the baboons and their relatives, including the drill and mandrill *Mandrillus,* the mangabeys *Cercocebus* and *Lophocebus,* the talapoin *Miopithecus,* and the swamp monkey *Allenopithecus,* as well as some species of *Macaca,* the females develop periodic (generally monthly) sexual swellings—prominent fluid-filled swellings of the vulva and perineum, sometimes extending to the anus and under the base of the tail—which are at their maximum around the time of ovulation. At this time they are most attractive to the males, and exhibit the behavior patterns known as estrus: sexual solicitation, and cooperation in mating.

▼ *The hamadryas baboon lives in rocky and arid areas in northern Ethiopia and southwestern Arabia. This group consists of a juvenile male and three females, with offspring. This is the usual foraging unit.*

N. Tanaka, Orion Press/Bruce Coleman Ltd

▲ The mountains of northern Honshu are snow-covered for more than half the year. In this harsh climate live Japanese macaques, also known as Japanese snow monkeys. During the winter they feed mainly on bark.

Macaques

The macaques, of the genus *Macaca,* occupy in Asia the same ecological niches that the baboons and their relatives occupy in Africa. Like savanna baboons, most macaques live in multi-male troops; some are largely terrestrial, others mainly arboreal; some develop sexual swellings in the females, others do not. The crab-eating or long-tailed macaque *Macaca fascicularis* is one of the smaller species, weighing about 5 to 8 kilograms (11 to 18 pounds); it is widespread in tropical Southeast Asia, from Vietnam and Burma to Java, Timor, Borneo, and the Philippines. The closely related rhesus monkey *M. mulatta* replaces it in monsoon and deciduous forest regions, extending from northern India into China, as far north as Beijing. Even less tropical is the Japanese macaque *M. fuscata,* which in northern Honshu lives in regions that are snow-covered for more than half the year. The so-called Barbary "ape" *M. sylvanus* lives in North Africa, and a small colony of them on the Rock of Gibraltar is kept going by fresh importations from Morocco every few years.

Colobid monkeys

The colobids have been separated from the cercopithecids since the Middle Miocene, about 12 million years ago, and the colobid digestive specializations are unique among primates. Whereas the cercopithecids are more diverse in Africa and have one genus in Asia, the colobids proliferate in Asia and are represented by only the colobus monkeys *Colobus* and *Procolobus* in Africa.

The mantled colobus *Colobus guereza* is a beautiful monkey with long silky black fur, varied with a white ring around the face, a white tail-tuft, and long white veil-like fringes along the sides of the body. It mostly lives in small one-male troops in forests in Kenya, Uganda, northern Tanzania, Ethiopia, and northern Zaire. The troops are territorial, the territories marked by the deep rattling roaring calls of the males. A study in the

Kibale Forest, Uganda, found that this species can survive for several months on little else but the dry, mature leaves of just one species of tree, the ironwood *Celtis durandii;* in Ethiopia it eats mainly *Podocarpus* leaves. A related species, the black colobus *C. satanas,* of the coastal forests of Cameroun, Gabon, and Equatorial Guinea, lives mainly on seeds (but of several tree species), avoiding the leaves which, in that region, are high in tannins and other toxic chemicals.

Colobus monkeys gallop, rather than walk, along branches and, at the end of the branch,

launch themselves into space without breaking stride and land in the neighboring tree, grabbing hold with their hook-like, thumbless hands.

In Asia, colobids are represented by the langurs; they differ from African colobus monkeys by having thumbs (though they are short), but their leaping skills are quite comparable. The entellus or sacred langur *Semnopithecus entellus* of India and Sri Lanka represents the monkey-god Hanuman of Hindu mythology, and consequently it is tolerated around Hindu temples and even in some towns, and is often fed by the faithful as a religious duty.

▲ *The adult male proboscis monkey (above, left) has a rather grotesque appearance. This species is confined to riverine and mangrove forests in Borneo. The golden snub-nosed monkey (above, center) is a rare, gaudily colored monkey from the cool mountain forests of southwestern China. The eastern black and white colobus or mantled guereza (above, right), once hunted for its beautiful skin, is now common again in light forest in East Central Africa. It is even found quite close to Nairobi.*

PRIMATE HANDS AND FEET

feet   hands

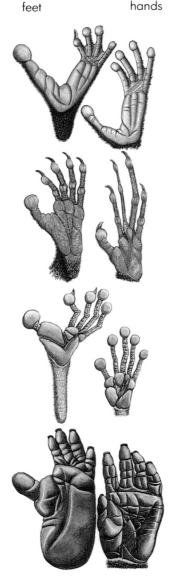

▲ The indri (top) clings to tree trunks; its great toe and thumb are stout and give a strong, wide grip. The aye-aye (second from top) has claws on all its digits except the great toe; it climbs by digging its claws into the bark, rather than by clinging like most other primates. The tarsier (third from top) has disc-like pads on its toes and fingers, to increase friction. The gorilla (bottom) has divergent thumb and great toe; the palm is broad to support its huge weight, and the great toe is stouter and less divergent than in other catarrhines, except humans.

▶ Muller's gibbon lives in Borneo. It is a close relative of the white-handed gibbon, from which it differs in color and in its longer, more trilling call. Gibbons have elongated arms and move mainly by brachiation — that is, they swing by their arms from branch to branch. They are smaller and have narrower chests and less mobile wrists than other apes.

Some langur troops are multi-male, others are one-male harems. In districts where one-male groups are usual, the surplus males form a bachelor band which, every two and a quarter years on average, invades a bisexual troop and ousts the troop male. The new troop male, one of the former bachelors, then tries to kill all the troop's unweaned infants. This unpleasant behavior trait appears on present evidence to be quite widespread among primates, especially in Old World monkeys, but is most conspicuous and best studied in these sacred langurs. Sarah Blaffer Hrdy, who has studied infanticide in langurs most intensively, points out that, when a female loses her infant, she stops lactating and so begins her sexual cycles again; the new troop can start breeding at once and so increase the reproductive output of the harem male before he, in turn, is ousted 27 months later.

Other langurs live in East and Southeast Asia. The forests of Malaysia and western Indonesia contain two genera, *Presbytis* and *Trachypithecus*. Mostly they live in the same areas, but whereas the silvery leaf-monkeys *Trachypithecus auratus* (of Java, Bali, and Lombok) and *T. cristatus* (from Sumatra, Borneo, and the Malay Peninsula) are relatively unvarying from place to place, the half dozen or so species of *Presbytis* are enormously diverse—every mountain bloc, offshore island, or forested region between two rivers has its own species or distinctive subspecies. But further north, where *Presbytis* does not occur, species of *Trachypithecus* may also show marked geographic variability. The capped langur *T. pileatus* of the Assam–Burma border region, and Francois's langur *T. francoisi* of the limestone hills of the Vietnam–Laos–Gwangxi border region, both vary enormously from one small area to another, and there is even a distinctive species, the beautiful golden langur *T. geei*, restricted to a small area in Bhutan.

In the mountains of Sichuan in China lives the bizarre and gaudily colored golden snub-nosed monkey *Rhinopithecus roxellana*. The males have fiery red-gold fur on the underside, flanks, and limbs, and around the face (the bare facial skin being pale blue), with long black fur on the back; they weigh 15 kilograms (33 pounds). The females are a drabber yellow-gold, again with black on the back, and weigh only 10 kilograms (22 pounds). Both sexes have curious, upturned, leaf-like noses. Two related species live in China, and another in Vietnam; a fifth species, the douc langur *Pygathrix nemaeus*, of southern Vietnam and Laos, has a flat nose but is even more gaudily colored.

The largest colobid is the brick-red-colored proboscis monkey *Nasalis larvatus* of Borneo. Males weigh up to 24 kilograms (53 pounds), females less than half that. The young have long, forward-pointing noses; in females these stop growing at maturity, but those of males carry on enlarging into enormous drooping Punch-like adornments.

(A nickname for them in Indonesia is *belanda*, which literally means "Dutchman"!) Proboscis monkeys live in troops in coastal areas, especially in and around mangroves, and along large rivers. They seem as much at home in the water as in the trees. It is one of nature's treats to see a huge male solemnly wade upright, arms held clear of the water, into a stream, lapsing into an energetic dog-paddle when he gets out of his depth.

Closely related to the proboscis monkey is the curious pig-tailed langur or simakobu *Nasalis concolor* (often placed in a separate genus, *Simias*) from the Mentawai Islands. Its nose is like a small version of the juvenile proboscis monkey's and it is the only colobid with a short tail.

APES: HAIRY AND NAKED

The apes, or hominoids (gibbon, orang utan, gorilla, chimpanzee, and human), differ from their sister-group, the Old World monkeys, in many ways. They have no tail: the few remaining caudal vertebrae, which make up the tail in other mammals, are variably fused together into a shelf-like small bone, the coccyx. The apes habitually sit (sometimes stand) upright, so it is no surprise that their lumbar vertebrae, being required to bear the weight of the upper body, instead of adding flexibility to the spine, are reduced in number (from seven or eight to four or five) and are much shortened and broadened. The thorax is broad—broader side-to-side than it is deep front-to-back—and the scapulae are on the back of the thorax, not on the sides, as in the Old World monkeys. The shoulders and wrists are very mobile. Apes, except for humans, have long arms, generally longer than their legs, but in fact in chimpanzees and gorillas it turns out that the arms are not especially long compared with the trunk—only in gibbons and orangs are the arms genuinely elongated.

There has been a lot of discussion about the evolutionary meaning of these anatomical features. Gibbons move mainly by brachiation, that is swinging by their arms under branches, and chimpanzees sometimes do this too, though orang utans clamber about with any old combination of arms and legs, and some forms of gorillas rarely climb at all and certainly never brachiate when they do. But, perhaps because of nineteenth century misunderstandings about the natural history of the apes, the idea that the hominoids' skeletal anatomy is "for brachiation" has become entrenched in both the specialist literature and the popular imagination, and it is almost impossible to get rid of. After all, did not our own ancestors "come down from the trees", where they had been swinging about by their arms? The facts suggest otherwise. First, gorillas are very largely terrestrial and chimpanzees are at least partly so—and these are our closest living relatives. Orang utans, the other great apes, do not brachiate. Gibbons (lesser

▲ *The siamang is the largest of the gibbons. It is seen here with an inflated throat sac, uttering the booming lead-in to its harsh, shrieking call.*

round to pull in the fruit. If brachiating ever was in our ancestry, it was surely only at a time when we shared a common ancestor with all other hominoids, and that common ancestor was small in size.

But even then the brachiating specializations do not go very far, for gibbons possess very many features, especially in the skeleton and muscles of their forelimbs, which are clearly valuable for their way of life but are not found in great apes or humans. When the human/great ape ancestor separated from the gibbon ancestor, it increased in size, and there was no more brachiation. When the human/gorilla/chimpanzee ancestor separated from the orang utan's ancestor, it came down from the trees. It is open-country existence, not terrestrial life as such, that distinguishes us from our closest relatives.

There are other hominoid characters too. An odd, and little-known, hominoid character is that we all possess a vermiform appendix, a thin projection on the end of the cecum (the blind gut at the junction of the small and large intestines), filled with lymphoid tissue. No other haplorhine primate possesses an appendix, so it must be an organ that the common ancestor of the hominoids developed, presumably to supplement the immune system. There is a popular belief that the human appendix is an evolutionary vestige (usually, it is suggested, of the cecum, and it is true that the cecum is very small in hominoids), but it is not.

Gibbons: arias and acrobatics in the tree tops

The smallest of the apes, the gibbons, are slenderly built, and have extraordinarily long arms and long narrow hands in which the thumb is deeply cleft from the palm, as far as the wrist; the legs are long, too, a fact often missed because the long arms overshadow everything else about them. They move easily and rapidly through the trees by brachiation: hand over hand over hand, in an up-and-down swinging gait, occasionally dropping easily to a lower branch, or jumping between trees.

All gibbons, as far as is known, live in pairs: a male and a female, with up to four offspring. The pair occupy a tree-top territory. From its center in the early morning, they launch into song, a duet: the female takes the major part, the male punctuating it at intervals with shorter, less coloratura phrases. The duet is primarily a spacing call, though by requiring the male and female to interweave their different contributions, it reinforces the pair-bond. Gibbons do not necessarily sing every morning, but when they do, the forest resounds with their soaring melodies as one pair after another takes up the theme, each from its own small territory. The wonders of gibbon acrobatics and operatics should not, however, obscure the fact that they are in fact only distantly related to the other hominoids.

The best-known species is the white-handed gibbon *Hylobates lar,* from northern Sumatra, the

apes), the ones that do brachiate, are not only small in size, but they also have the least hominoid anatomical specializations: the narrowest chests, the longest lumbars, and the least mobile wrists.

Observing apes in the wild suggests what the so-called "brachiating characteristics" are really all about. Great apes are lazy: they like to sit in amongst their favorite food plants and rake them in without moving much; to do this they sit upright and reach out, rotating their shoulders and wrists to pull in the vegetation or fruit. For a big-bodied animal it is a very worthwhile economy to be able to feed almost without moving. Gibbons practise a variation on this: they hang from the ends of branches, or sit on them, and again simply reach all

Malay Peninsula, Thailand, and parts of Burma and Yunnan. Weighing 5 to 7 kilograms (11 to 15 pounds), these small gibbons may be either buff or dark brown-black in either sex (the exact shade differs according to subspecies), both color types being found in the same population. They are remarkable, too, for their extremely dense fur, with more than 1,700 hairs per square centimeter (11,000 per square inch) of skin on the back—two or three times that in most monkeys. Related species live in Java, Borneo, southern Sumatra, and Cambodia: they differ in color and in vocalizations. Kloss's gibbon *H. klossii*, from the Mentawai Islands, which is completely black and has much less dense fur, is one of the few species in which the sexes sing separately and do not duet. In Burma, and extending into Assam and Bangladesh, is found the larger hoolock gibbon *H. hoolock*, which weighs up to 8 kilograms (18 pounds). In this species, both sexes are black while juvenile (with a white brow-band), but on maturity—at about 7 years, as in all gibbons—the female turns buff-brown. The hoolock gibbon lives in monsoonal semi-deciduous forests, not rainforests like the *lar*-group, and it has a small inflatable throat-sac that gives it a much harsher call.

The concolor gibbon *H. concolor* lives in the Vietnam–Yunnan border region, and on Hainan Island. With age, the color changes in the female from black to buff, just as in the hoolock, although the two are not very closely related; the male, in addition, has an upstanding tuft of hair on the crown. Only the male has a throat sac; his voice is little more than a harsh shriek, while the female is melodious like members of the *lar*-group. Further south in Vietnam and Laos are species in which the black form has whitish cheek whiskers.

The largest gibbon is the siamang *H. syndactylus*, which weighs 9 to 12.5 kilograms (20 to 27½ pounds). Both sexes are black and have very large throat sacs, so that their calls (quite deafening at close quarters) are harsh barks and shrieks, punctuated by the boom of the throat sacs being inflated. They live in Sumatra and West Malaysia, in the same forests as *H. lar* and *H. agilis*. Compared with these smaller gibbons, siamangs eat more leaves and less fruit, vocalize less frequently, have smaller territories, and seem more closely pair-bonded. Remarkably, recalling those other pair-bonded primates, the marmosets, the adult male carries and babysits the infant, at least beyond its first year.

▼ *White-handed gibbons, the best known of all gibbons, are still quite common in the forests of the Malay Peninsula, and will survive as long as the forests survive. But how long will that be?*

Jean-Paul Ferrero/Auscape International

Silvestris/Australasian Nature Transparencies

▲ The orang utan, the "man of the woods", is the least humanoid of the great apes but is still a remarkably intelligent and adaptable creature.

Alain Compost/Bruce Coleman Ltd

▶ Each night, the solitary orang utan makes a fresh sleeping nest high in the trees. Leaves and branches are pulled into a platform shape in the crown of a tree.

Orang utan: dignified tree-top mandarin

The orang utan *Pongo pygmaeus* is today restricted to Borneo and northern Sumatra. On Borneo it is still locally common in lowland forests, but in Sumatra, where it is not found south of about Lake Toba, it ascends to considerable altitudes. The name "orang utan" means simply "wild person", or "man of the woods", in Malay and Indonesian; indigenous people call it *maias* (in many Dayak languages) or *mawas* (in Sumatra).

Orangs are covered in sparse red hair through which the rough blue-grey skin can be seen. Bornean orangs *Pongo pygmaeus pygmaeus* tend to have thinner, more maroon hair, whereas in Sumatran orangs *P. p. abelii* the hair is lighter, more gingery, and fleecier, and often very long, especially on the arms. An orang's arms are very long, the legs appearing short by comparison; both the hands and feet are powerful and curved, with shortened thumb and great toe, and long second to fifth digits.

Orangs are sexually mature at about 7 years of age, after which time females more or less cease growing; adult females weigh 33 to 42 kilograms (73 to 92 pounds), and are 107 to 120 centimeters

(42 to 47 inches) high when standing bipedally. But the males carry on growing until they are 13 to 15 years old; a fully mature male weighs 80 to 91 kilograms (176 to 200 pounds)—though in zoos they become obese and can weigh much more—and stands normally 136 to 141 centimeters (53½ to 55½ inches) high, though a giant of 156 centimeters (61½ inches) has been recorded. In their teens, too, the males develop fleshy cheek-flanges; in Sumatran males, these tend to be smaller, and protrude sideways from the face, whereas those of a Bornean male are often very large and swing forward as he moves.

The red ape lives a solitary life. Females have overlapping home ranges of about 2 square kilometers (¾ square mile). A mature male occupies a much larger area, of 8 square kilometers (3 square miles) or more, which is not exactly a territory, as the area is too large to exclude others from it, but other males will try to avoid the resident. From time to time the resident utters a succession of deep carrying roars: other males move away from the sound, whereas females may gravitate towards it. At night, every individual builds a nest, high in the trees; vegetation is packed securely into the crown of a tree, with a special rim placed around the main platform. When the orang lies in the nest, it makes a cover by pulling down other vegetation. Generally a fresh nest is made each night, even if the animal has not traveled far that day.

More than half the diet is fruit, with leaves, bark, and chance morsels like insects making up the rest. Rainforest trees come into fruit rather irregularly, depending not on season as such but on a combination of factors such as recent and current patterns of rainfall, cloudiness, temperature, wind, and so on. It has been pointed out that for a very large animal it is far more economical to be able to remember the locations of favorite fruiting trees over a wide area, and to be able to calculate the likelihood of a given tree being in fruit, than to go searching. And so, the theory goes, the high intelligence of the great apes is an almost inevitable consequence of being large and eating fruit.

There has been much discussion as to whether female orangs have restricted estrus, or are willing to copulate throughout their sexual cycle of 28 days. It now appears that they have only short estrus periods: for a few days around the time of ovulation, the female actively seeks out and solicits a mature, flanged male, and they associate together for a few days. But subadult males, not sought out by females, will chase and rape females at any time. It has been suggested that, as the penis is so short, a copulation will fail unless the female cooperates, so that most infants are born out of consortships with mature males, not from rapes by subadults. The gestation averages 245 days.

In the past, orangs were much in demand by zoos, and as they were obtained by shooting mothers and collecting their infants (most of

John Cancalosi/Bruce Coleman Ltd

which would then die from inadequate substitute care), a dozen orangs would die for every one that reached a zoo. Now this is banned by international agreement: orangs breed well under good zoo management, and an increasing proportion of zoo orangs are now captive-born. The new threat to the orang is deforestation; even where, as in Indonesia, there is government concern for conservation, and logging is no longer regarded as the inevitable fate of every forest, the inexorable increase of the human population creates land hunger, and more and more inroads are made into reserves and national parks by cultivation and by deliberately lit fires, which have been especially devastating during the long periods of El Niño drought.

▲ A mature male Sumatran orang utan. Sumatran males differ from Bornean in their flatter cheek-flanges, the smattering of downy hair on the face, and their long beards and moustaches.

D. Parer & E. Parer-Cook/Auscape International

D. Parer & E. Parer-Cook/Auscape International

▲ *A silverback male mountain gorilla reclines in the dense ground vegetation of the Virunga Volcanoes.*

▼ *A young mountain gorilla at play.*

Gorilla: Africa's gentle giant

The gorilla *Gorilla gorilla* is the largest living primate. Many people have a false impression of just how large a gorilla is. Reputed half ton, nine foot high monsters just do not exist: the record for an adult male in the wild is 219 kilograms (482 pounds) and 195 centimeters (77 inches), and even this is most unusual. The average adult male weighs on average 175 kilograms (385 pounds) and stands 156 centimeters (61½ inches) bipedally; a female weighs 85 kilograms (187 pounds) and stands 137 centimeters (54 inches) high. Like orangs, gorillas reach sexual maturity at about 7 years. At this time females cease growing, but males grow until they are about 12 or 13, and at full physical maturity develop a silvery-white "saddle" on the back, contrasting with the black hair (hence, a fully mature male is known as a silverback). The prominent brow ridges, smooth black skin, and flaring nostrils contrast with the high forehead, rough grey skin, and tiny nose of the orang. Gorillas have a pungent odor, emanating from the armpits: like humans, and also chimpanzees, they have a large cluster of specialized sweat glands there.

Gorillas are very largely terrestrial. They do climb, but not too commonly (especially not the big silverbacks), and they always travel long distances on the ground. They travel on all fours, with their hands not palm-down on the ground but flexed, so that the weight rests on the middle joint of the fingers: this is called knuckle-walking. Like orangs, they make nests every night—not such complex, well-made nests, and usually on the ground, though in a few areas there seem to be "traditions" of building nests in trees.

A gorilla troop consists of a silverback male (sometimes two, or even more), a few blackback (or subadult) males, and several females and young. The troop wanders over a large home range of 10 to 20 square kilometers (4 to 8 square miles) which overlaps that of other troops, with whom relations are normally peaceful. When a male matures, he has the option of staying in the troop, or leaving it. It has been suggested that he will stay in his natal troop if there are any females in it who are potential mates for him, otherwise he will leave and live alone for a while, sometimes following other troops and trying to "kidnap" females from them, usually by invading a troop when its silverback is off-guard and rounding up one or more females. In the fracas that ensues, there is often a fight, and infants may be killed—perhaps deliberately by the invading male, for an infant seems to be a bond between a female and her troop. Females may also take the opportunity to leave a weak male during an encounter of this nature, but if the new male in turn proves weak and cannot attract further females to him, then the females he already has will take the first opportunity to go elsewhere. The most stable kind of troop is probably one in which there is a second silverback ready to take over.

The gorilla has a fragmented distribution within Africa. There are three subspecies. The western lowland gorilla *G. g. gorilla* lives in West Central Africa. The much blacker eastern lowland gorilla (*G. g. graueri*) lives in Zaire, east of the Lualaba River, extending from the lowlands near the river up into the mountains. The black, long-haired, large-toothed mountain gorilla *G. g. beringei* lives in the Virunga Volcanoes, on the Zaire–Rwanda–Uganda border, and in the Bwindi Forest in south-western Uganda. It has recently been suggested that eastern gorillas should be separated as a full species, *Gorilla beringei*, and that Bwindi Forest gorillas should form a different subspecies of it.

Within lowland rainforest, gorillas seem to favor secondary growth; in the mountainous parts of their range they live in open forest and in bamboo thicket, up to 3,500 meters (11,500 feet). They feed in montane areas on bamboo shoots and stems of tall herbs, saplings, and small trees, but in lowland areas more fruit is eaten, especially tough, woody fruits. This diet seems to have its drawbacks: 17 percent of gorilla skulls in museums have dental abscesses, and most mountain gorilla skulls show severe breakdown of the bone between the teeth—perhaps because of the packing of bamboo fibers, although other bamboo-eating gorilla populations seem not to show this condition. And 17 percent of the skeletons show traces of spinal arthritis, not surprising for such a heavy, partially upright animal.

▼ Despite years of civil turmoil in Rwanda, the population of mountain gorillas in the Virunga Volcanoes is slowly recovering from near-extinction. The proportion of young is growing and is now nearly 40 percent.

Yann Arthus-Bertrand/Auscape International

Helmut Albrecht/Bruce Coleman Ltd

▲ *Common chimpanzees spend much of their time on the ground, but are also active and efficient climbers.*

Chimpanzees: cunning and charisma

The chimpanzee, like the gorilla, is a specialized knuckle walker; like the gorilla it is black, has large brow-ridges and a smooth skin which is black, at least in the adult. But most analyses, whether biochemical or morphological, tend to indicate that the chimpanzee is even more closely related to humans than it is to the gorilla. So perhaps this indicates that the characteristics it has in common with the gorilla were those of our own distant ancestor too.

The common chimpanzee *Pan troglodytes* is still widespread in West and Central Africa, reaching as far east as western Uganda and the extreme west of Tanzania. It lives not only in rainforest but also in montane forest, open forest, and even savanna woodland where it has not been exterminated. The very big Central African subspecies *P. t. troglodytes,* often called the bald chimpanzee, is very black-skinned from quite an early age, and adults go bald on the crown—males in a triangle, narrowing back from the forehead, females totally. Males weigh 60 kilograms (130 pounds) on average and average 120 centimeters (47 inches) in height when standing bipedally; females average 47.5 kilograms

(105 pounds) but are nearly as tall as the males, so they are more lightly built. The East African or long-haired chimpanzee *P. t. schweinfurthii* is much smaller: males in the famous Gombe Stream population of Tanzania average only 43 kilograms (95 pounds) and females 33 kilograms (73 pounds). They are less intensely black than Central African ones, have a more marked beard and cheek whiskers, and they do not go bald so completely or so early. The West African or masked chimpanzee *P. t. verus* is small, like the East African; it is very dark around the eyes and nasal bridge, but less so elsewhere on the face; it develops a long gray beard with maturity, does not go very bald, and when young has a central parting on the head.

But much more characteristically different, and generally considered a distinct species, is the pygmy chimpanzee or bonbon *Pan paniscus,* from south of the great bend of the Zaire River. Despite its name, it is about the same weight as the two smaller forms of the common chimpanzee, and stands 119 centimeters (46½ inches) high; it has longer legs, shorter arms, and a smaller head than the common chimpanzee, as well as long side-whiskers, rather small brow-ridges, and a face that is black from birth except for the contrastingly pinkish white lips.

Chimpanzees spend about half their time on the ground; in the trees, they knuckle-walk along branches, or brachiate beneath them. They feed on fruit, and to a lesser extent on leaves, bark, insects, and even on vertebrate prey; some populations very regularly kill monkeys and other medium-sized mammals, which they hunt cooperatively (males do most of the hunting). At night they make nests in trees, like orangs, though their nests are not as complex; and, like gorillas, nests may also be made in the heat of the day for "siesta".

Chimpanzees live in large groups, which have been called communities by some authors and unit-groups by others: these number anything from 20 to 100 or more, but mostly its members associate in small parties, of varying composition, either bisexual parties roaming widely, searching for fruiting trees, or more sedentary nursery groups of mothers and infants. The community as a whole occupies a territory, whose boundaries are patrolled by the adult males against incursions by outsiders. The well-known field researcher Jane Goodall recorded an instance, in the Gombe National Park, of short-lived warfare between two neighboring communities, which ended in the virtual extermination of the smaller one.

Adult males seem almost always to remain in the communities in which they were born, whereas females seem always to leave them and join neighboring communities, sometimes changing communities more than once during their lives.

This pattern is clear in the *schweinfurthii* populations of Tanzania and Uganda, which have been well studied, but probably also applies to other common chimpanzees and to pygmy chimpanzees. Pygmy chimpanzees, however, appear to travel in groups of more consistent composition. In the East African chimpanzees males seem to form close bonds of association (perhaps because of their relatedness?). In pygmy chimpanzees females also form close ties, which appear to involve homosexual interaction.

There is a well-marked difference between the two species in their heterosexual behavior. Females in both species have sexual swellings: enormous pink doughnut-shaped excrescences involving vulva, perineum, and anus. In the common chimpanzee, the swelling begins to appear after menstruation and gradually enlarges, reaching a maximum at the time of ovulation, after which it rapidly detumesces; the total cycle length is 35 days. The female solicits mating around the time of maximum swelling, though she will mate at other times: about 75 percent of copulations occur in the week or 10 days when she is fully swollen. In the pygmy chimpanzee, however, the cycle length averages 46 days in the wild (though in two captive females it was only 36 days); she

has large sexual swellings for about half of this time, and never completely lacks them except when lactating. Most primatologists studying pygmy chimpanzees have found no restriction of copulation to any phase of the sexual cycle; only one has even reported a tendency for sexual activity to diminish during the times of lowest swelling. Moreover, common chimpanzees, like almost all other non-human primates (gibbons and orangs mating while hanging from trees may form an exception), mate dorso-ventrally, but pygmy chimpanzees very frequently mate ventro-ventrally, maintaining eye contact all the while.

When a female common chimpanzee has her maximum sexual swelling, males gather round her like flies and she willingly mates many times, apparently indiscriminately. But she may instead choose to form a consortship with one particular male; the pair slip away for several days and sometimes the female will consort with the same male next time she comes into estrus. It is claimed that relatively more infants are born as a result of consortships than after promiscuous behavior. Pygmy chimpanzees also form consortships.

Gestation lasts about 230 days, shorter than the orang (245 days) or the gorilla (267 days). Like all great apes, lactation goes on for a long time—two or three years, or even more—and there is an interval of four years between live births. Sexual maturity is not reached until about 8 years of age and, unlike other apes, males as well as females seem to stop growing at this time. Because, like many other primates, including humans, there is a period when sexual cycles are non-ovulatory ("adolescent sterility"), a female chimpanzee may be 12 or more before she gives birth for the first time. Chimpanzees, like other apes, might live 35 to 40 years, sometimes more in captivity, and very aged females seem to show signs of menopause.

▼ Tool-use, even tool-making, is known in many populations of wild chimpanzees. Here a youngster, still not very proficient, pokes a stick into rotten wood to extract grubs.

Peter Davey/Bruce Coleman ltd

modify twigs and grass stems the better to obtain termites, ants, or honey (in different populations); in one population of chimpanzees, in Ivory Coast, they select special hammer-stones to open *Cola* nuts, and leave them in known places, beside tree buttresses, which make suitable anvils. Alone of all mammals except humans, great apes readily (and spontaneously, given a few days exposure) come to recognize their own images in mirrors: they have, that is, a concept of self. Other animals—monkeys, gibbons, elephants—have the intelligence to use a mirror's reflection to find hidden food, and can recognize other individuals reflected in it, but never themselves.

Most staggering, perhaps, have been the language experiments. Using hand signs (based on

Dr Ronald H. Cohn/The Gorilla Foundation

▲ *All three great apes can be trained to use symbols to communicate, in a very rudimentary way, with their trainers, even sometimes with each other. Here Koko, a female gorilla, inspects a machine.*

Almost human?

It has long been known that great apes are more intelligent than any other animal (I almost forgot to add, "except humans"). They learn faster; they can learn a greater variety of special skills; and they can learn more complex tasks. They very clearly figure out how to perform the tasks set them in psychological tests, sometimes by apparent flashes of insight, sometimes, it would seem, by careful logic. They can learn to use implements, too. Gorillas are not so good at this, although they are very patient when faced with complex machinery, but chimpanzees and orang utans readily learn to modify objects to make tools, for example, to obtain food. All this has been known since the work of R.M. Yerkes in the 1920s, but the degree to which apes' minds resemble our own was never suspected then. It is now known that some groups of chimpanzees make tools in the wild: they

THE EVOLUTIONARY TANGLE

Gibbon (*Hylobates*), orang utan (*Pongo*), gorilla (*Gorilla*), chimpanzee (*Pan*), and human (*Homo*) are the five living genera of the Hominoidea. Perhaps more work has been undertaken to try to elucidate how these five genera are interrelated than on the whole of the rest of the primates put together. After all, we are involved; human pride is at stake.

Human chauvinists would prefer that we were the odd man and woman out (in other words, the sister-group of the rest); until 25 years ago, it was customary to place all the great apes together in one family, the Pongidae, separate from humans, Hominidae. Even many of those who can come to terms with scientific findings, and bring themselves to accept the reality of evolution, would like to see us separated from the other primates altogether. In the nineteenth century it was usual to place humans in a separate order, Bimana, from the other primates, which were called Quadrumana. In the 1920s and 1930s the great Australian anatomist and zoologist Frederic Wood Jones proposed his "tarsier hypothesis"—that our ancestor was a small long-legged beast related to the tarsier and so separate from the monkey/ape lineage since the Oligocene (over 25 million years ago), if not earlier.

In the 1950s a famous paper, "The Riddle of Man's Ancestry" by W. L. Straus, an American anatomist, proposed that the characters we share with the great apes are more or less illusory, and that we are at least as different from them as are the gibbons; his reasoning convinced many eminent authorities, and there was no surprise when, in the early sixties, the paleontologist Elwyn Simons claimed that a Miocene fossil ape, *Ramapithecus* (known only from fragmentary jaws!), was a human ancestor contemporary with *Dryopithecus*, the putative ancestor of living great apes—hence 20 million years at least must be envisaged for the time of separation of the two lines.

American Sign Language for the deaf, or Ameslan), chimpanzees learned to symbolize objects and actions, adjectives and sometimes fairly abstract concepts; in later experiments, both orangs and gorillas have learned the same. In the first flush of enthusiasm, many psychologists claimed that their apes could produce sentences with full syntax, but this now seems dubious. A continuing experiment at the Yerkes Primate Center in Atlanta, Georgia, uses computer consoles: chimpanzees learn to press different buttons for different words, and communicate in this fashion with their trainers, with "untrained" people (thus eliminating the danger of the "Clever Hans" effect, in which the trainers' unconscious body cues influence the animal's behavior)—and, to a degree, with each

Biochemists such as Morris Goodman and Vincent Sarich began working on this problem in the early 1960s. While disagreeing on details, they all agreed that the blood proteins (albumin, transferrin, hemoglobin, and others) of the gibbons were most different from those of other hominoids, those of the orang utan next; while those of humans, chimpanzees and gorillas were very close to each other (some of their proteins being nearly or quite indistinguishable). This created an impasse: paleontologists insisting that they were working with the actual record of evolution, biochemists pointing out that the fossil material was incomplete.

The Gordian knot was cut in 1982 by Peter Andrews and John Cronin, who showed in a classic paper that, first, a fossil ape, *Sivapithecus,* which had previously been confused with *Dryopithecus,* is actually quite distinct; second, the by-now quite well-represented cranial remains of *Sivapithecus* are exactly what one would expect in the primitive ancestor of the orang utan; and third, the jaws assigned to *Ramapithecus* fall within the range of variation of *Sivapithecus.* The supposed special resemblances of *Ramapithecus* to humans (such as the thickness of the enamel on the molar teeth) turn out to be shared by *Sivapithecus,* too, and seem to be in any case just primitive features, retained from the common ancestor of humans and all great apes. This massive re-evaluation of the fossil data meant that there was no longer any evidence for an early separation of the human lineage; on the contrary, there was now good evidence for the early separation of the orang utan lineage, in agreement with the biochemical findings.

More recently, analysis of the morphological characters of the living hominoids has confirmed the picture deduced long ago by Goodman and Sarich: the gibbon ancestors split off first, then the orang ancestor, while gorillas, chimpanzees, and humans are very closely related to one another, while the gorilla is the sister-group of the chimpanzee plus human group, as we now know.

The Gorilla Foundation/National Geographic Society

▲ *Koko with her pet kitten, All Ball. Koko was devoted to the kitten and, reportedly, grieved intensely when it was hit by a car and died.*

other. Two young chimpanzees, for example, acquired the facility of requesting each other to share different types of tools for obtaining food. It has been claimed that pygmy chimpanzees do best of all at this kind of intellectual activity; so far there has been no independent test of this claim, but it is recorded that a pair of recently captured pygmy chimpanzees *spontaneously* made iconic hand-signs to each other in the context of their favorite activity, ventro-ventral copulation.

All this has, as might be expected, produced a great deal of soul-searching among biomedical researchers—people who would have no compunction about using monkeys for medical or "curiosity" experimentation—about whether it is ethically justifiable to use apes for such purposes. Because orangs and gorillas are classified as endangered species, they are not used in any case, but until recently chimpanzees were not considered endangered and were used for terminal experiments, disabling surgery, or injection with infective agents without a second thought. This has all changed. Sadly, chimpanzees are now also put in the endangered category as their wild populations have declined catastrophically in the last 20 or 30 years, and only captive-bred chimpanzees may now legally be used in endangering research, at least in the United States; but it is very noticeable how thinking on caging, husbandry, and some types of research has been altered by the cognitive studies of the past twenty years. As one physiologist has put it, we have discovered that apes are close to us mentally as well as physically, and this carries ethical obligations with it.

COLIN GROVES

Rod Williams/Bruce Coleman Ltd

► The only nocturnal monkey is the douroucouli, or night monkey, widespread in the forests of South and Central America. It has enormous eyes, but no eye-shine.

Rod Williams/Bruce Coleman Ltd

▲ The smallest living monkey, the pygmy marmoset, lives in the Upper Amazon rainforests, where it feeds partly on tree sap and gum, which it extracts by gouging into the bark with its incisor teeth.

SPECIAL EYES

Primates all have large eyes whose visual fields overlap, so that they have stereoscopic vision. Most primates also have color vision. These two features are not restricted to primates, but there is a complex sorting of the neuroanatomy of the visual system in the brain, which is shared between primates and just two other groups of mammals, the flying lemurs or colugos, and the fruit bats.

Lining the inside of the eyeball is the retina, where incoming light is intercepted. The retina has two types of receptor: rods and cones. Rods are capable of receiving very low-level light; cones interpret color. Day-living primates, as might be predicted, have a high proportion of cones; in the human retina there are 7 million cones and 125 million rods, which is indeed a high proportion (though about average for diurnal primates). Night-living primates have little use for color vision and therefore have nearly all rods—dwarf lemurs of genus *Cheirogaleus,* for example, have 1,000 rods for every cone.

Many nocturnal or crepuscular mammals have eye-shine: when a bright light is shone into their eyes, a reflection comes back. The reflection is due to a crystalline layer behind the retina, called the tapetum lucidum. This acts to increase the amount of light that passes across the retina, and so it assists night vision.

Most strepsirrhines are nocturnal, whereas nearly all haplorrhines are diurnal; but all strepsirrhines (even the diurnal ones) have a tapetum. With this eye-shine, strepsirrhines seem basically adapted to lead nocturnal lives, and no monkey can compete with them at this, so in mainland Africa and Asia, where they share their range with monkeys, there is a clear split: all strepsirrhines are nocturnal, all monkeys are diurnal. But in Madagascar, where the lemurs have it all to themselves, some have become diurnal and fill a "monkey-like" niche. And, interestingly, in Latin America, where there are no strepsirrhines, one genus of monkeys, *Aotus,* has become nocturnal, filling a "lemur-like" niche (the other nocturnal haplorrhine is the tarsier, of Southeast Asia). But no haplorrhine has a tapetum, so *Aotus* and *Tarsius* compensate by have the largest eyes of all primates, and apparently they are the only ones to have no cones in their retinas.

Haplorrhines have a special feature of the retina which strepsirrhines lack: a macula, or cones-only region, with a depression in its center. This is an area of exceptionally high visual acuity. The two nocturnal haplorrhines, *Aotus* and *Tarsius,* have some of the histological features of the macula, even though they have no cones.

Thus, primates retain in their eyes the stamp of their evolutionary past: all strepsirrhines have a tapetum, whether they are nocturnal or not, because their common ancestor was nocturnal; all haplorrhines have a macula, whether they are diurnal or not, because their ancestor was diurnal.

▼ *The diademed sifaka of eastern Madagascar (below, left) moves and feeds by day, but the structure of the retina discloses a nocturnal ancestry.*

▼ *Photographed at night, the lesser bushbaby's eyes (like those of all strepsirrhines) reflect the light of the flash camera.*

Jane Burton/Bruce Coleman Ltd

Frans Lanting/Minden Pictures

CARNIVORES

ORDER CARNIVORA
- 11 families • 106 genera
- 270 species

SMALLEST & LARGEST

Least weasel *Mustela nivalis*
Head–body length: 15–20 cm
(6–8 in)
Tail length: 3–4 cm (1–1½ in)
Weight: 100 g (3½ oz)

LAND Polar bear *Ursus maritimus*
Head–body length: 2.5–3 m
(8–10 ft)
Weight: 800 kg (1,750 lb) +

AQUATIC Southern elephant seal
Mirounga leonina
Head–tail length: 4.9 m (16 ft)
Weight: 2.4 tonnes (5,300 lb)

CONSERVATION WATCH
!!! The red wolf *Canis rufus*;
Ethiopian wolf *Canis simensis*;
Mediterranean monk seal
Monachus monachus; and *Viverra
civettina* are listed as critically
endangered.
!! 24 species are listed as
endangered, including: African
wild dog *Lycaon pictus*; Iberian
lynx *Lynx pardinus*; tiger *Panthera
tigris*; snow leopard *Uncia uncia*;
giant-striped mongoose *Galidictis
grandidieri*; Liberian mongoose
Liberiictis kuhni; marine otter
Lutra felina; Colombian weasel
Mustela felipei; European mink
Mustela lutreola; Steller's sealion
Eumetopias jubatus; Hawaiian
monk seal *Monachus schauinslandi*;
Cozumel Island coati *Nasua
nelsoni*; Cozumel Island raccoon
Procyon pygmaeus; giant panda
Ailuropoda melanoleuca; lesser
panda *Ailurus fulgens*; otter civet
Cynogale bennettii; falanouc
Eupleres goudotii; crested genet
Genetta cristata.
! 37 species are listed as
vulnerable, including: cheetah
Acinonyx jubatus; clouded leopard
Neofelis nebulosa; lion *Panthera
leo*; wolverine *Gulo gulo*; giant
otter *Pteronura brasiliensis*;
Galapagos fur seal *Arctocephalus
galapagoensis*; northern fur seal
Callorhinus ursinus; sloth bear
Melursus ursinus; spectacled bear
Tremarctos ornatus; Asiatic black
bear *Ursus thibetanus*; fossa
Cryptoprocta ferox.

From our fascination with man-eating tigers to our close association with domestic cats and dogs, beasts of prey have always been considered very special by humans, even though they represent only about 11 percent of all mammals. Carnivores have been seen historically as vicious predators, high on the food chain, and often as competing with humans. Within the order, however, are animals that live entirely on plants, invertebrates, large mammals, and even fruit. Nonetheless, within the Carnivora are some of the strongest and most formidable of all mammals.

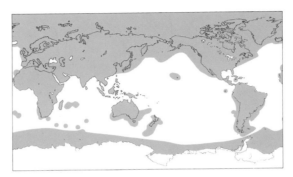

FASCINATING MAMMALS

In spite of our special fascination, we know relatively little about most of the wide diversity of species represented in the Carnivora. From the smallest, the least weasel *Mustela nivalis* at a mere 100 grams (3½ ounces), to the largest, the southern elephant seal at 2.4 tonnes (5,300 pounds), our knowledge of the biology of the group is scant. Within the group is the fastest land mammal, the cheetah *Acinonyx jubatus*, and perhaps some of the slowest: the earless seals (Phocidae), which are better adapted for the sea than the land. Some carnivores are extremely common where they occur, for example the North American coyote *Canis latrans* or the small Indian mongoose *Herpestes auropunctatus*; others, however, are quite rare, known only from a handful of museum specimens, such as the Columbian mountain weasel *Mustela felipei*.

Carnivores are found on every continent except Antarctica and in nearly every type of habitat, with species found in the oceans, the Arctic, tropical rainforests, prairies, temperate forests, deserts, high mountains, and even the urban environment. They are rarely a common or abundant part of any ecosystem, usually because they are found near the top of the food chain. They are the only order of mammals that have ocean-going, terrestrial, arboreal, and semi-fossorial species. Among the mammals, including known fossil groups, at least ten orders of mammals are primarily insectivorous and twenty are herbivorous, but only one modern and one extinct order are carnivorous: the Carnivora and the Creodonta. With such a wide diversity of kinds of mammals, one might ask why

these animals are grouped together and what they have in common.

WHAT IS A CARNIVORE?

Some carnivores are not classified in the order Carnivora; some members of the order Carnivora are not carnivores; and some carnivorous mammals are neither carnivores nor in the order Carnivora! So, just what is a carnivore? A carnivore is an animal that eats principally meat, but the word is most often used as a general term to refer to the members of the Carnivora. Being carnivorous means that an animal specializes in eating meat as the principal part of its diet; this is a food habit of some, but not all, of the members of the Carnivora. The order Carnivora consists of a unique group of mammals that all share a common evolutionary history, that is, the members of the Carnivora all share a particular set of ancestors. Members of each family within the order are also united by sharing a particular set of common ancestors, and these ancestors can be identified by certain unique morphological features. The first members of the Carnivora had adaptations that allowed them to eat meat more efficiently than other competing mammals around at the time (creodonts). These features are retained in nearly all of the present members of the order, despite the fact that many animals have modified them and now feed on many things besides meat. Some species specialize in eating fruit, insects, bamboo, worms, nuts, berries, fish, crustaceans, and seeds. Although these animals may specialize in a particular kind of food, for example, the giant panda *Ailuropoda melanoleuca* feeds on bamboo, nearly all are considered opportunists, that is, they will not turn down an easy meal when it is available. Consequently, many members of the Carnivora are probably more appropriately called omnivores—they eat a wide variety of plant and animal foods—and the portion of meat in their diet may range from nearly all, such as a cat, to hardly any, such as the giant panda.

HOW ARE CARNIVORES DIFFERENT?

Throughout their common evolutionary history, members of the Carnivora have acquired some unique features that distinguish them from other orders of mammals. These features relate to the basic eating habits of the early members of the order and can be grouped under these modifications. In order to survive by eating other animals, carnivores must not only be able to eat them efficiently, but must first be able to catch them. Modifications to find the animals they feed on, such as acute eyesight, hearing, or smelling, allow them to locate prey before it locates them. Second, speed and dexterity are essential, and modifications have allowed carnivores to catch prey quickly without undue expenditure of energy. Third, once the prey is caught, a carnivore must be able to kill and digest the animal efficiently and derive all its essential food requirements from it. And fourth, the carnivore must be smarter than the animal it is trying to catch!

The most widely accepted theory about the relationships among carnivore families is based primarily on modifications in the ear region and the sensitivity to hearing particular frequencies. Carnivores have a highly developed ear region, often with more than one inner ear chamber, which increases the sensitivity to certain frequencies and makes it easier to locate prey that makes those particular sounds. Skulls from each of the families of carnivores can be identified from their unique ear region alone.

Carnivores have adapted a variety of ways for feeding on everything from large mammals to fruit.

The single most important feature, and the one that is most often cited as uniting the entire order, is a unique modification of the teeth for eating meat. Carnivores have the fourth upper premolar and the first lower molar modified to form two vertical, sharp cutting surfaces, which slide against each other in a manner similar to the blades of a pair of scissors. At this location on the skull the jaws have their greatest force, making this pair of teeth, called the carnassial pair, extremely efficient as shears. However, some carnivores have evolved into non-meat-eaters and they have modified the carnassial pair to best utilize different food resources. Plant-eating and fruit-eating Carnivora have lost most of the vertical shear surface and increased the horizontal crushing surface (similar to humans). Insect-eating Carnivora have small teeth with small shearing surfaces that can pierce the exoskeleton of insects and get to the food inside.

THE DOG-LIKE CARNIVORES

The dog-like carnivores of the suborder Caniformia are believed to have evolved from an ancestor similar to the extinct family Miacidae, which lived during the Eocene. Caniforms are mostly terrestrial, although two families, the Phocidae and Otariidae (discussed at the end of this chapter), have become exclusively aquatic. The caniforms are the more numerous of the two suborders, both in terms of biomass and number of species, and they are found in all continents and oceans. The

CAT JAW SHOWING CARNASSIAL SHEAR

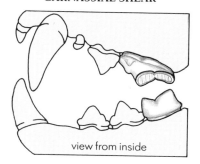

view from inside

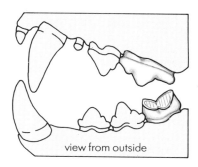

view from outside

▲ Shading indicates the shearing facets on an upper premolar and a lower molar. (Cats lack the first upper and first two lower premolars.)

▼ Silver-backed jackals and a spotted hyena feeding on a wildebeest, killed by hyenas.

terrestrial forms share a uniquely formed internal ear, different from that found in the feliforms. This group includes the largest (elephant seals) and the smallest (least weasel) of all carnivores.

Dogs, foxes, and jackals

The family Canidae is dominated by the genus *Canis*. The family is usually divided into roughly four groups: the foxes of the genera *Alopex*, *Urocyon*, and *Vulpes*; the dogs or wolves of the genera *Canis*, *Lycaon*, and *Cuon*; the interesting South American canids; and the more insectivorous canids, the raccoon dog *Nyctereutes procyonoides* and the bat-eared fox *Otocyon megalotis*, which are quite different from all other members of the family.

Canids live in open grasslands and show adaptations for the fast pursuit of prey. They originated in North America during the Eocene (57 to 37 million years ago) and are now distributed throughout the world. The familiar domestic dog

and all of its breeds and varieties is believed to have evolved from a wolf-like ancestor.

Canids rely on hearing and smell to locate their prey. They are opportunists and highly adaptable, with a generalized dentition and digestive system that allow them to shift their diet from eating nothing but fruit at certain times of the year to being entirely carnivorous. Canids eat all types of vertebrates, insects, fruit, mollusks, and carrion. Some species, such as the wolf, African hunting dog, and red dhole *Cuon alpinus*, use cooperative hunting, which allows them to prey upon animals many times their own body size. Others, like the bat-eared fox, are insectivorous.

Canids have a long face and usually two molars in each jaw. The carnassial shearing teeth have the shearing surface at the front end of the tooth and then a crushing surface at the heel, usually with two cusps.

Canids are terrestrial carnivores with long, bushy tails, long legs, and slender bodies. Only one, the North American gray fox *Urocyon cinereoargenteus,* can climb trees easily. Many modifications reflect the pursuit way of life of canids, for example, the bones of the forelegs are interlocked to prevent accidental rotation during running. All canids have large pointed ears. The skin is almost completely without sweat glands and there is a dorsal scent gland at the base of the tail.

Some canids have had a long association with humans, and many cultures have a rich history of legends and folktales associated with these animals. Wolves are viewed as raiders of livestock and will attack humans when threatened. Coyotes may sometimes be seen around the edges of several urban communities; they are despised by landowners, who see them as predators of small livestock. The fox's attributes are legendary and the "fox raiding the henhouse" has even become a cliché. For these and other reasons massive predator control programs have been initiated in several areas and, combined with habitat destruction, have led to the demise of many species of canids from a large portion of their original range. Canids are economically important and are of high value in the fur industry. The raising of foxes on farms has in most cases eliminated the need for extensive hunting to supply the trade.

Foxes The foxes are small canids with acutely pointed skulls and bushy tails. The desert foxes are the lightest colored and the smallest, and have proportionately the largest ears. The fennec fox

▶▼ *Members of the Canid family. Above right, the bush dog of South America. Below: top, maned wolf of South America; middle, red wolf of North America (an endangered species); bottom, the African hunting dog.*

Vulpes zerda (500 grams to 1.5 kilograms, 1 to 3 pounds weight) has an almost comical appearance, with enormous ears that help to dissipate heat and to locate prey. All of the foxes have basically the same eating habits: they eat whatever small vertebrates are in the area where they occur, but they also depend heavily on other food items such as insects and fruits. The Arctic fox *Alopex lagopus* (3 to 5 kilograms, 6½ to 11 pounds weight) occurs in two color phases, a brown form and a steel-gray or blue form. Both color types change to pure white in the winter. Most foxes are considered solitary hunters: cooperative foraging does not give much advantage to a predator that depends on small vertebrates such as mice. Some recent natural history studies indicate, however, that fox society may be much more complex. In some areas the basic social unit is usually the breeding pair and their offspring; however, recent studies suggest that in other areas the social unit consists of small groups, usually composed of one adult male and three to six females or vixens. The group appears to hunt in a specific area, although the individuals will disperse while foraging.

Dogs and wolves The true dogs and wolves are the most carnivorous of the canid family, with the African hunting dog being the only pure carnivore. Within this group are also the most social species. Wolf packs, for instance, have a well-developed hierarchy centering on the dominant breeding pair, known as the alpha male and female. Cooperative hunting by six to twelve individuals, seen in the wolf, Indian dhole, and the African hunting dog, allows the predator to specialize in very large prey such as zebra, antelope, elk, and deer. Despite the image of wolves running down moose in deep snow, or Indian dholes disembowelling a large deer, their success rate is actually quite low, almost always below 10 percent, and they usually have to go for long periods without eating. Coyotes and the four species of jackals (*Canis aureus, C. mesomelas, C. simensis,* and *C. adustus*) are smaller, more opportunistic foragers, and they concentrate more on medium-sized to smaller vertebrates. Coyotes can and will eat almost anything. Whereas the specialized, highly carnivorous *Canis* species are threatened by predator control and habitat destruction, coyotes do not seem to be suffering as

much from human intervention and they are increasing their range throughout North America.

South American canids The three kinds of South American canids are the long-legged maned wolf *Chysocyon brachyurus,* the "foxes" of the genus *Dusicyon,* which are morphologically and ecologically somewhere between the Northern Hemisphere *Canis* and *Vulpes,* and the unusual bush dog *Speothos venaticus.* The bush dog has the most carnivorous dentition of the group, but it is a most unusual-looking canid, with short legs, a rotund body, small ears, and a short tail. Its head and body length ranges from 55 to 75 centimeters (22 to 30 inches) and it weighs 5 to 7 kilograms (11 to 15 pounds). It will run in packs of five to ten individuals and feed on the larger capybaras. The maned wolf, so named because of its erectile dorsal crest of hair, has long, stilt-like legs adapted for tall grass prairie. It is a predator of medium-sized mammals such as rabbits, pacas, and armadillos.

Bat-eared fox This is one of the most peculiar of the canids, and the only one that is entirely insectivorous. It is unique among the Carnivora in having four to eight small extra molars, providing more chewing surfaces for feeding on insects. Like the fennec, it has relatively large ears that help both in locating prey and dissipating heat.

Raccoon dog Named because of its black face mask (similar to the North American raccoon), the racoon dog is the only canid that does not have any kind of bark. It is also distinguished from the rest of the family by living in dense undergrowth of forests. Raccoon dogs are commercially valuable in the fur industry and have been introduced into many areas.

▼ *More members of the Canid family. Top, gray fox of North America; bottom left, bat-eared fox of Africa; bottom right, raccoon dog of Asia.*

John Shaw/NHPA

Bears and pandas

The largest terrestrial carnivore, and the only bear that is exclusively a meat eater, is the polar bear. Males may weigh over 800 kilograms (1,750 pounds). Despite their Hollywood image, however, most bears do not feed regularly on large prey, and they are primarily herbivores or insectivores, although, like all carnivores, they are opportunistic and will take advantage of a good meal when they see it. The giant panda (75 to 160 kilograms, 165 to 250 pounds weight) is almost exclusively a bamboo feeder, whereas the sloth bear *Melursus ursinus* has long claws and powerful front teeth for breaking logs and getting insects and other invertebrates. Its long snout and long tongue are typical of insectivorous mammals. Another adaptation for insect eating is that it has lost its two upper middle incisors, providing a convenient channel for its long, prehensile tongue.

The ancestor to modern bears first appeared in the Oligocene, a red-panda-sized animal in the genus *Cephalogale*. At this time, all of the bear ancestors had long tails and were of this size. Throughout their evolutionary history, bears have increased in size, becoming the largest of the land-dwelling carnivores, with large massive heads, stocky bodies, and short tails (except for the red panda). Eurasia is believed to have been the center of the evolution of ursids.

There is a great similarity in general body form in all the modern bears save one, the panda. The anatomical features that define this family

▲ *Young polar bears at play. Largest of the terrestrial carnivores and well adapted to life in the Arctic, this species feeds almost entirely upon seals.*

▼ *The Asian black bear is an excellent climber.*

Jean-Paul Ferrero/Auscape International

principally concern changes in dentition from the shearing–cutting form typical of carnivores, to a crushing–pulverizing form with broad, cuspidate molars. Bears have the largest molars of any carnivores. All the shearing function of the carnassials has been lost. Indeed, the first few premolars, typically shearing teeth, are rudimentary and often drop out at an early age.

Except for the giant and red pandas, the bears are all basically single colored, with some species showing striking color differences within the species. Nearly all show marks on the chest.

The members of this family are terrestrial and semi-arboreal omnivores; only the polar bear is semi-aquatic. Three species—red panda, sloth bear, and sun bear—are nocturnal, and the others are diurnal. They have delayed implantation, giving them a relatively long gestation. The northern temperate species are absent from their natural range for up to six or seven months of the year. During the winter their natural foods are not available, so they store up body fat just before retreating into a den or cave and go into a winter dormancy, distinguished from true hibernation because there is no drop in body temperature. During this period they do not eat and live entirely off their body fat. It is also during this time that one to five extremely small young—they weigh only about 1 percent of the mother's weight—are born in a protected, warm environment.

Brown and polar bears Also known as the grizzly, Kodiak, and Kamchatkan bear, the brown bear *Ursus arctos,* with a head–body length up to 3 meters (10 feet) and a weight of up to 450 kilograms (990 pounds) is one of the largest ursids. Along with the Asian black bear *Ursus thibetanus* and American black bear *Ursus americanus,* it feeds on tubers, berries, fish, and carrion. Living in an extremely harsh environment without such food resources, polar bears feed almost exclusively on marine mammals, especially seals. Polar bears can reach 3 meters (10 feet) long and weigh up to 650 kilograms (1,430 pounds). In search of seals, they may travel on ice floes up to 80 kilometers (50 miles) a day. The females do not breed until they reach five years of age. When pregnant, they isolate themselves in a den to have the young.

Pandas The giant panda and the red panda are the only carnivores that feed almost solely on bamboo. The giant panda was originally known as the "black and white bear" and Dwight Davis's classic morphological study in 1964 confirmed its position as a member of the bear family. The position of the red panda has been less clearcut. Because of a superficial resemblance to the North American raccoon, some have proposed placing it in the raccoon family, but morphological studies do not support this. Molecular studies seem to point to a position between the bear and raccoon families, and some suggest it is the surviving member of an extinct group of carnivores.

Philippa Scott/NHPA

▲ The giant panda feeds exclusively on bamboo.

▼ The red panda eats bamboo, other vegetation, and insects.

Joe van Wormer/Bruce Coleman Ltd

▲ *The common raccoon of Central and North America has adjusted to life in suburbs and farmland.*

Raccoons and coatis

Raccoons The inquisitive raccoon *Procyon lotor* (5 to 15 kilograms, 11 to 33 pounds, weight) is familiar to most people in North America. It is, however, just one of the more highly successful members of the procyonid family. This family is centered in the tropics of Central and South America, with only the raccoon, which can often be found associated with humans, increasing its range into northern North America. The family Procyonidae is entirely restricted to the New World. Although all of the members of this family are omnivores, some have become quite specialized and concentrate on eating fruit to an

▼ *The coati of tropical America used its flexible snout to expose the insects and tubers that form much of its diet.*

extent not seen in many other carnivores. All are nocturnal, except for the diurnal coatis.

The family is defined principally on morphological features. Among these are features of the skull, such as a deep posterior-oriented pocket in the external ear region, and the presence of inner ear sinuses, which are small pockets connected to the inner ear region where the ear bones lie. The carnassial shear, one of the basic carnivore traits, has been highly modified in this family. Only the ringtail has retained a sharp cutting edge, while the other species in the family have modified dentition with blunt cusps or even no cusps at all, such as the kinkajou *Potos flavus*. Except for the kinkajou and the olingo *Bassaricyon gabbii,* procyonids have a distinctive black facial mask and rings on their tail.

Raccoons are mainly omnivorous, but they prefer aquatic prey, such as frogs, fish, and crustaceans, and will also feed on various nuts, seeds, fruits, and acorns. They have the peculiar habit of washing their food with their highly dexterous forepaws. They are nocturnal, solitary, and found mainly in forested areas, where they live in dens in hollow trees or rock crevices. For this reason, the cutting down of old trees by many landowners has dramatically reduced raccoon populations.

Coatis The closely related coatis are diurnal inhabitants of woodlands and forests. Dramatic individual color variation is a common feature in coatis. One litter described in 1826 had a red, a gray, and a brown individual in the same litter;

black coatis and red coatis are not infrequent. Coatis are omnivorous, their food chiefly composed of insects and roots. Females and juveniles live in bands ranging from four or five to fifty individuals; they forage by spreading out over the forest floor and moving slowly along, digging, rooting, and investigating all possible sources of food. Their highly flexible snout allows coatis to forage in crevices and holes. Their tail acts as a balancing organ when the animal is in the trees.

Males are not allowed in the group except during the mating period. One dominant male will work its way into the foraging group and then mate with the females. Soon after mating, the males are expelled from the group. Adult females separate from the group to give birth.

Little is known about the behavior of the mountain coati *Nasuella olivacea,* apart from the fact that they are believed to feed predominantly on grubs and worms.

Olingos and kinkajous These are almost entirely arboreal and are mainly fruit eaters, but they may also feed on some birds and small mammals. With an elongate body form and uniformly colored fur, they have the general appearance of a primate. They are fast-moving, extremely agile and active, and travel constantly during the night. Kinkajous are 40 to 60 centimeters (16 to 24 inches) in length and weigh 1.5 to 2.5 kilograms (3 to 5½ pounds). The kinkajou was originally thought to be a lemur, and the manner in which it uses its feet and its prehensile tail, its arboreal habits, and its fruit diet may still lead observers to associate it with monkeys rather than carnivores. Kinkajous are the only New World carnivore with a prehensile tail.

Weasels, otters, skunks, and badgers

With nearly double the number of species of any other carnivore family, the family Mustelidae is clearly the most successful evolutionary group. They are found in every type of habitat, including both salt and fresh water, and on every continent except Australia and Antarctica. Despite their widespread and common occurrence, they are rarely seen by humans, because of their nocturnal, arboreal, or burrowing habits. The feature for which the family is most famous (or infamous!) is the scent gland located around the anus that can secrete a pungent, foul-smelling odor, sometimes for great distances.

The reproduction biology of the group is unique among the Carnivora: the males tend to be considerably larger than the females (5 to 25 percent) and the sexes live separately for most of their lives. Mating does not occur by mutual consent: the males violently accost the females, hold them down, often with a neck bite, and copulate vigorously for long periods of time. The stimulation of the copulation process causes the female to release an egg, which ensures successful

fertilization. In other mammals, after fertilization the egg implants on the uterine wall and begins to develop. In mustelids, the egg "floats" in the uterus and implantation is delayed for many months.

This family of fur-bearing animals has played an important historical role in the exploration of many areas. The furs of otters, minks *Mustela vison,* weasels, wolverines *Gulo gulo,* and martens of the genus *Martes* still bring high prices, and the trapping of these species is tightly controlled.

Mustelids are distinguished from other carnivores by their musk gland: two modified skin glands near the anus. They have five toes on each foot with non-retractile claws. The skull is low and flat, with a very short face. The teeth have no second upper molar and no distinctive notch in the fourth upper premolar—features found in all other carnivores. Most mustelids have an hour-glass shaped upper molar.

This diversified family has recently been reevaluated. The subfamily Mustelinae (weasels, wolverines, martens) appear to be a natural group. Molecular studies have questioned the position of the skunks (subfamily Mephitinae) within the Mustelidae, and suggested that they belong in a separate family. The badgers (subfamily Melinae) have several problematic species, with some showing similarities to the skunks, and the North American badger (Taxidea) being so different that some suggest placing it in its own subfamily.

▲ The kinkajou lives in tropical American forests, feeding mainly upon fruits and nectar. It is one of the very few carnivores to have a prehensile tail, which is a great aid to climbing.

▼ One of the smallest members of the raccoon family, the ringtail is also one of the most carnivorous. It preys upon lizards and small mammals but also eats fruits and nuts.

Erwin & Peggy Bauer/Bruce Coleman Ltd

▲ An arboreal member of the weasel family, the South American tayra feeds mainly upon fruit. Its long tail is not prehensile.

▼ The wolverine (also known as the glutton) is a thickly furred carnivore that inhabits the colder parts of Eurasia and North America.

Weasels, wolverines, and martens The mustelines are one of the most successful groups of carnivores in the world. The terrestrial and semi-aquatic members of this family, referred to as weasels, have short round ears, extremely long cylindrical bodies, and short legs. They feed mostly underground and at night, chasing their prey, usually rodents, directly into their burrow systems. One species, the mink, is semi-aquatic and spends a large amount of its time foraging in the water for mollusks, crustaceans, and fish. The black-footed ferret *Mustela nigripes* was a specialized feeder on North American prairie dogs, but as landowners eliminated the prairie dogs when they converted their land to pasture for cattle, the ferret also became extinct in the wild. Martens are arboreal weasels, and their body size ranges are slightly larger. The largest marten, the fisher *Martes pennanti,* ranges in length from 45 to 75 centimeters (18 to 30 inches) and weighs 2 to 5 kilograms (4½ to 11 pounds). It is the only carnivore to feed heavily on porcupines. The marten accomplishes this by speed and dexterity. The porcupine's main defense is to keep its back, with the erect quills, toward the predator. The fisher will quickly move to the head, where there are no quills, and attack. Once the porcupine is dead, the marten flips the animal over and feeds from its underside. One South American arboreal marten, the tayra *Eira barbara,* depends on fruit for a large portion of its diet. The largest mustelid is the wolverine, which is 1,000 times heavier than the smallest species, the least weasel. Wolverines have robust bodies and short legs, and feed on small ungulates and medium-sized vertebrates. They are adept at killing relatively large animals in deep snow. Their winter fur, which has hollow hairs, is prized by Eskimos for its insulative properties.

Badgers Badgers are generally stocky, short-legged, large mustelids with powerful jaws that

Konrad Wothe/Bruce Coleman Ltd

Jeff Foott/Bruce Coleman Ltd

interlock and support a crushing dentition. Most are solitary, except for the Eurasian badger *Meles meles*, which lives in social groups and feeds mostly on earthworms and grubs. With a head and body length of 60 to 85 centimeters (24 to 34 inches), and weight of 10 to 15 kilograms (22 to 33 pounds), the Eurasian badger is slightly larger than the North American badger *Taxidea taxus*. The latter feeds on medium-sized vertebrates and uses its powerful front claws and limbs to burrow after rodents. Little is known about the natural history of the Southeast Asian badgers.

Otters The most highly successful aquatic carnivores, apart from the seals, are the otters, which spend nearly all of their time foraging in water for food. They occur in both fresh and salt water, and have webbed feet and stiff whiskers which they use as tactile sensors. Most otters are typical mustelids, leading solitary lives, but the sea otter is gregarious, and the males and females will form separate groups in different coastal areas. The giant otter *Pteronura brasiliensis* which ranges in head–body length from 85 to 140 centimeters (34 to 55 inches) and in weight from 20 to 35 kilograms (40 to 75 pounds), will sometimes forage in groups. The otter's muscular tail provides most of the force for swimming, and it attacks its prey with its teeth. The oriental short-clawed otter *Aonyx cinerea* uses its feet to grasp invertebrates to feed on. The sea otter often uses "tools", such as a flat rock, to break open mollusk shells.

Skunks The New World skunk subfamily, the Mephitinae, is perhaps the most infamous of the mustelids. The anal gland, present in all mustelids, is here modified to an organ that can project a foul-smelling liquid for a great distance. The skunk's coloration of vivid black and white patterns is a warning to most potential predators of what is to come. Skunks are omnivores, feeding on insects and occasionally small vertebrates, occupying a niche not unlike that of some mongooses.

Jen & Des Bartlett/Bruce Coleman Ltd

▲ *Sea otter and young. This North Pacific species spends almost all of its life at sea, even sleeping on the surface.*

▼ *Striped skunk, swimming. Although not a habitual swimmer, the skunk—like most mammals—can swim when necessary.*

▲ *The banded linsang, from the Indonesian region, is an arboreal civet that feeds on small invertebrates.*

▼ *The common genet, from the Mediterranean region, is a versatile predator on small vertebrates.*

THE CAT-LIKE CARNIVORES

The cat-like carnivores, of the suborder Feliformia, are believed to have evolved from an ancestor similar to the extinct family Viverravidae (which is also included in this suborder); these were more civet-like and lived during the Eocene. Feliforms are terrestrial; only the otter civet and the marsh mongoose *Atilax paludinosus* are semi-aquatic.

Civets and genets

An important commercial product for many centuries, civet oil is obtained from the civet gland of certain members of the Viverridae. This gland, located in the genital region, is unique to this family of carnivores. Civet oil, or just simply "civet", is refined and used as the basis for perfume.

Included within the Viverridae are some of the most diverse and interesting species of the Carnivora, but they are the least known and least studied. This possibly is because most occur in tropical regions, and are nocturnal and solitary, making observations difficult. Viverrids are found throughout Africa and southern Asia. Some species are quite common, like the common palm civet, sometimes called the toddy cat, because of its love for fermented palm sap (toddy). The common palm civet is also a notorious raider of coffee plantations. With the exception of a few species, we know little about the relative abundance or distribution of this family, our knowledge is limited to scattered records accumulated over the last century. With the destruction of the tropics occurring at such an alarming rate, there is the real danger that one of the most common components of the tropical rainforest, the family Viverridae, may suffer even more than others.

With the viverrids' occupation of such a wide diversity of ecological types, it is difficult to characterize the morphology of the whole family. The most common feature, the civet gland, is found in all genera except a few, although its exact

shape and position varies. Most viverrids have a very generalized dentition, with rather long faces, and a tail as long or sometimes longer than the head–body length. The ear region is diagnostic for this carnivore family: viverrids have an inner ear divided into two chambers, externally visible on the skull.

Because of the wide diversity in this family, it is useful to group its members by subfamilies. Here we will consider four: the palm civets (Paradoxurinae), Malagasy civets (Cryptoproctinae), true civets (Viverrinae), and banded palm civets (Hemigalinae).

True civets The most primitive subfamily, and the one with the most species, is the Viverrinae. Viverrines provide the basis for the perfume industry and are represented in Europe by the common genet *Genetta genetta*. The members of the diverse genus *Genetta* are arboreal, long-nosed cat-like animals that are found throughout Africa and into southern Europe. They prey on small vertebrates and invertebrates. The largest civet, the African civet, and the corresponding Indian civets, of the genera *Viverra* and *Viverricula,* are terrestrial —one of the few non-arboreal groups—and prey on medium-sized to small vertebrates.

Malagasy civets These civets share little in overall appearance. The fossa, the dominant carnivore on Madagascar, is extremely cat-like, and occupies a cat-like niche on the island. At 70 to 80 centimeters (28 to 32 inches) in length and 7 to 20 kilograms (15 to 45 pounds) in weight, it is about the same size as the African civet, making them the largest viverrids. The fossa occupies a niche similar to that of the clouded leopard of Southeast Asia. Fanalokas are rather fox-like in their ecology and appearance. Perhaps one of the most interesting of the Madagascar civets is the falanouc, a small (2 to 5 kilogram, 4½ to 11 pound), long-nosed animal with a bushy tail like a tree squirrel where it stores fat reserves. The falanouc and fanaloka are the only carnivores that produce young that are active immediately after birth.

Palm civets The palm civets demonstrate the evolution of a group of meat-eating carnivores into fruit-eating. The palm civets of Southeast Asia spend nearly all of their time in the tree canopy, and one species has been documented as feeding on over 30 species of fruit. Recently, the masked palm civet *Paguma larvata* has become a popular food item in restaurants in China, causing alarm over its conservation status and future.

Banded palm civets Like the Malagasy civets, the banded palm civets are a rather loose array of morphological types. This group is the least studied of the viverrids and occurs only in Southeast Asia. The banded palm civet appears to eat small vertebrates and insects. Both Owston's civet *Chrotogale owstoni* and the falanouc have long narrow snouts and feed on earthworms and other invertebrates. Both of these civets have strikingly

Frans Lanting/Minden Pictures

beautiful, broad transverse bands on their back. The otter civet resembles an otter and lives in the same habitat as an otter, but nothing is known of its natural history.

▲ *The fossa is the largest carnivore of Madagascar, an island that has no indigenous members of the dog or cat family.*

▲ *The African civet is the largest civet.*

Frans Lanting/Minden Pictures

▲ The ring-tailed mongoose hunts in trees for small reptiles, mammals, and birds. It is one of five species, restricted to Madagascar, that are distinct from all other mongooses.

▼ Although related to the carnivorous hyenas, the aardwolf feeds almost exclusively on termites. Its teeth are degenerate but it is far less specialized for its diet than other mammalian "anteaters"

Mongooses

Mongooses, small carnivores of the family Herpestidae, have long slender bodies, well adapted for chasing animals down into burrows, though they mostly eat insects. Most mongooses are diurnal; none possesses the retractile claws necessary for adept tree climbing; and only a few species have external markings. Mongooses are basically a uniform color, though a few have stripes on the neck or different tail coloration. The banded mongoose *Mungos mungo* has black transverse bands on its body, and some Malagasy mongooses (Galidiinae) have longitudinal dark stripes running the length of their bodies. Mongooses are often the most abundant carnivore in an area, and they are found from deserts to tropical rainforests, from southern Africa through the Middle East, India,

and Southeast Asia, northward to Central China. Famous as "ratters", mongooses have been introduced to many islands in the Old and New World. The diurnal mongoose was introduced to many Caribbean islands to control the nocturnal rodents feeding on sugar cane, but they were not a great success, and they are now considered a pest species in most areas where they were introduced.

Mongooses are small, ranging from about 25 to 60 centimeters (10 to 24 inches) in head and body length, with short faces and two molars in each jaw. Their inner ear is divided into two chambers, but, unlike the other feliforms, these chambers are located one in front of the other, not in the typical side-by-side arrangement. All mongooses have an anal sac that contains at least two glandular openings. This is best developed in the African mongooses and least developed in the Malagasy mongooses. The glands deposit scent with the feces, which communicates the sexual condition and other characteristics to other mongooses.

Mongooses attain sexual maturity at 18 to 24 months. Litter size varies from species to species, but generally from two to eight young are born after a six to nine month gestation. The adult female solely raises the young in the solitary species; in the more social mongooses, the young may be cared for by several adult females.

Mongooses are essentially terrestrial and are poorly adapted for climbing, only the African slender mongoose *Herpestes sanguineus* and the ring-tailed mongoose *Galidia elegans* frequent the trees. One species, the marsh mongoose, is semi-aquatic and another, the crab-eating mongoose *Herpestes urva,* feeds heavily on aquatic crustaceans. Most are solitary or live in pairs. Members of one subfamily, the Mungotinae, are noted for their colonies and gregarious social systems. The dwarf mongoose *Helogale parvula* and banded mongooses will forage in groups of 5 to 20, and colonies of the yellow mongoose *Cynictis penicillata* have been found with 50 individuals. Both the yellow mongoose and the meerkat *Suricata suricata* sometimes live in burrow systems closely adjacent to the Cape African ground squirrel of the genus *Xerus*.

Hyenas and aardwolves

Few who have heard the cry of the spotted hyena *Crocuta crocuta* in the still air of the African night will forget the presence of one of the most fearsome predators in Africa. The hyenas at one time dominated the carnivore niche, and fossil hyenas have been found in North America, Europe, Asia, and Africa. Now the entire family consists of four species representing three genera. Hyenas are essentially carnivores of open habitats and are unknown in forested regions.

The common names are descriptive: the spotted, striped (*Hyaena hyaena*), and brown (*Hyaena brunnea*) are not easily confused. They range in size

Wardene Weisser/Ardea London

from 1 to 1.4 meters (40 to 55 inches) in head and body length and from 30 to 80 kilograms (65 to 175 pounds) in weight. The wolf-like or dog-like appearance of the aardwolf *Proteles cristatus,* which also has stripes, is clearly different. All members have a distinctive dorsal mane and they possess an anal pouch which is structurally unique in the feliforms and is used to deposit scent. The skull is distinguished from those of all other families of carnivores in the inner ear region. The inner ear consists of two separate chambers, which are also found in the cats, civets, and mongooses, but in the hyenas these chambers, rather than being arranged side by side, as in other carnivore families, are located one on top of the other.

The members of the two genera of true hyenas, *Crocuta* and *Hyaena,* are amongst the largest predators known: they have massive, long forelegs and short hindlimbs. The remaining member of this family, the aardwolf, although it retains many similarities to the hyenas, feeds almost entirely on termites and has a much more slender build and peg-like teeth.

True hyenas The true hyenas have large heads with impressive high bony crests on the top of the skull, which serve as the attachment for a massive jaw muscle that gives hyenas one of the strongest bites of any carnivore. With robust, bone-crushing teeth, these predators are best adapted for feeding on carrion, and they are superb scavengers. Hyenas, like other carnivores, are opportunists, but they derive a large portion of their diet from scavenging on large ungulate kills. Their digestive system crushes and dissolves nearly all of the scavenged kill, but they regurgitate hooves, antlers, and other matter that cannot be dissolved into pellets. The spotted hyena participates in group hunting of anywhere from 10 to 30 individuals, and different types of hunting strategies are employed for different types of prey. The females of the spotted hyena are larger than the males and have one to four young.

Aardwolf The aardwolf is a solitary forager, and for most of the year it depends on a few species of termites. Aardwolves are nocturnal; they use many dens and locate their prey primarily by sound. Teeth are relatively useless to an animal that feeds almost exclusively on termites, and their dentition is so reduced that the teeth are not much more than small stubs. They measure 63 to 75 centimeters (25 to 30 inches) in length and weigh 7 to 12 kilograms (15½ to 26½ pounds).

Lions, tigers, and cats

Of all carnivores, cats, of the family Felidae, are clearly the most carnivorous. Cats cannot crush food: their teeth consist almost entirely of sharp, scissor-like blades, with no flat surfaces for crushing. To masticate food, cats must turn their head to the side and use their sharp, rasp-like tongue to manipulate food around their carnassials

to cut it into swallowable pieces. The distinguishing features of the felids are the horny papillae on the tongue, and the presence of only one molar in each jaw. The felids have short faces, forward-pointing eyes, and high domed heads. They do not show the diversity of body form prevalent in other carnivore families. Cats' binocular vision, unique dentition, and other modifications have been successfully adapted to a wide variety of environments, from the deserts of the Sahara to the Arctic. Although they are principally meat eaters, at least two also rely on other types of food: the fishing cat *Felis viverrina,* which depends heavily on crustaceans, and the flat-headed cat *Felis planiceps,* which will eat fruit.

All cats have some capacity to climb trees. They are commonly stalk or ambush killers, and the final capture is achieved by a sudden pounce.

Traditionally, the cats have been divided into three groups: the large cats, called the pantherines, the small cats, referred to as the felines, and the cheetah. The principal difference, besides size, between the pantherines and the felines is that the small cats can purr but not roar, and the large cats can roar, but not purr.

Cheetah The cheetah, which ranges in weight from 35 to 65 kilograms (77 to 143 pounds), is the only true pursuit predator in the cat family, and its speed is famous: it has been clocked at 95 kilometers per hour (60 miles per hour). Unfortunately, during this burst of speed, the cat expends tremendous amounts of energy and builds

▼ Famed as the fastest land animal, the cheetah is a sprinter rather than a long-distance runner: a chase seldom lasts more than 20 seconds.

Anup & Manoj Shah/Planet Earth Pictures

A. Visage/Auscape International

▲ *Tigers have a predominantly tropical distribution but the habitat of the Siberian race is covered in deep snow during the winter.*

Jonathan Scott/Planet Earth Pictures

▲ *African lions, the most sociable of the big cats.*

up massive amounts of heat. Generally, a chase will only last for about 20 seconds. After a short time, the heat load becomes so great that, like a runner who has run too hard, the cat overheats and finds itself out of breath. After capturing its prey, often the cheetah must stop, cool down, and catch its breath, and it is during this time that other predators, taking advantage of the cheetah's vulnerability, will steal its prey.

Tiger The pantherine cats contain some of the most impressive carnivores, such as the tiger *Panthera tigris,* the lion, and the leopard *Panthera pardus.* The tiger, the largest cat (130 to 300 kilograms, 280 to 660 pounds), is the only carnivore that will occasionally supplement its diet by attacking humans. It feeds on very large prey—50 to 200 kilograms (110 to 440 pounds)— hunting alone and waiting in ambush to strike its prey. It is a solitary cat; the main type of social unit is the adult female and her immediate offspring. Females and males occupy home ranges that do not overlap with those of others of the same sex.

Lion By contrast, the lion is the most social member of the cat family and shows the most marked sexual dimorphism: the males have impressive manes and are 20 to 35 percent larger than the females. The main social unit of lions is the pride, usually consisting of five to fifteen adult females and their offspring, and one to six adult males. Lions, which range in weight from 120 to

▲ Small cats. Upper left, the caracal of Africa and Asia; upper right, the clouded leopard of Asia; lower left, the ocelot of the Americas; lower right, the jaguarundi of the Americas.

240 kilograms (260 to 520 pounds), hunt cooperatively and regularly kill prey above 250 kilograms (550 pounds). The large males, with their highly visibly manes, do little of the actual killing. The normal hunting strategy is for the adult females to fan out and surround the prey and slowly stalk the intended victim; the kill results from a final, short burst of speed from an ambush.

Leopard The extremely widespread leopard, and its ecological counterpart the jaguar *Panthera onca,* show a surprisingly high amount of melanism. The leopard ranges in weight from 30 to 70 kilograms (65 to 155 pounds), the jaguar, 55 to 110 kilograms (120 to 240 pounds). They are solitary nocturnal hunters that seek smaller prey than the lions or tigers, and they are adept tree climbers. The clouded leopard *Neofelis nebulosa* (15 to 35 kilograms, 33 to 77 pounds weight) hunts almost entirely in the trees, feeding on monkeys, small aboreal mammals, and birds. Snow leopards *Panthera unica* (25 to 75 kilograms, 55 to 165 pounds weight) are uniquely adapted for high altitude, and snow-covered areas; they have feet well covered with hair. They are solitary predators of large ungulates.

W. CHRIS WOZENCRAFT

Stefan Meyers/Ardea London

▲ Leopards often rest in trees and may wedge dead prey in a fork to keep it out of reach of competitors and carrion-eaters.

▲ *A colony of Californian sealions. Sealions, which swim with their fin-like arms, hunt independently but form dense associations when they come ashore to breed.*

PINNIPEDS: LIFE ON LAND AND AT SEA

Seals, sealions, and walruses show adaptations to life in the water, with their streamlined bodies and limbs modified to form flippers. They spend much time in the water, but return to land to breed. Different modifications of the hind flippers make some more efficient movers on land than others. Members of the family Phocidae are not able to bend their hind flippers forwards at the ankle, so their progression on land is a humping movement. Phocids also lack an external ear pinna, and the nails on the hind flippers are all the same size. Phocids may be marine, estuarine, or fresh-water.

The members of the two families Otariidae (sealions and fur seals), and Odobenidae (walruses) are all marine and are able to bend their flippers forwards at the ankle, so their movement

on land is much more "four-footed". The nails on the hind flippers are very small on the two outer digits, but longer on the middle three.

Sealions and fur seals have a small external ear pinna, and slim cartilaginous extensions to each digit of the hind flippers which can be folded back so the middle three nails can be used for grooming. Walruses lack the external ear pinna, and the extensions to the hind digits are much smaller. Many skull characters also distinguish the three families of aquatic carnivores.

Interesting names are applied to pinnipeds. A large group is a herd, but a breeding group is a rookery. Adult males are bulls; females are cows; and the newborn young are pups (calves for walrus). Between about 4 months and 1 year the young is described as a yearling; a group of pups is a pod; and immature males are bachelors.

Most seals eat fish and squid, octopus, crustaceans of all sizes, including krill, mollusks, and lampreys. Leopard seals *Hydrurga leptonyx* and some sealions are not averse to penguins and carrion, and have been known to attack the pups of other seals. The diet, however, varies with geographic location.

Enemies vary according to locality and range from sharks, killer whales, and polar bears to humans and their pollution.

Sealions and fur seals

Both sealions and fur seals (otariids) and the walrus came from ancestors that lived in the North Pacific about 23 million years ago; these ancestors, in turn, came from early bear-like ancestors.

Sealions The five species of sealions are still primarily animals of the shores of the Pacific. They are big animals: the adult males range from 2 to 3 meters (80 to 118 inches) long; the females are smaller. Pups at birth are 70 to 100 centimeters (27 to 40 inches) long. Underfur hairs are few, so the coat is coarse, and on adult males the heavy muscular neck has a mane. The coat colour is a darkish brown in males, often lighter to gray in females, and in the male Australian sealion *Neophoca cinerea* the top of the head is whitish. Steller's sealion *Eumetopias jubatus* is the largest; the Californian sealion *Zalophus californianus* is the noisiest, barking almost continuously, and is the seal most frequently seen in zoos and circuses. The Australian sealion is restricted to Australian coastal waters, and Hooker's sealion *Phocarctos hookeri* is the most southerly, found on the Auckland Islands.

Fur seals There are two genera: *Arctocephalus*, with eight species, and *Callorhinus*, with one.

Callorhinus, the northern or Pribilof fur seal, has its main breeding area on the Pribilof Islands and Commander Islands in the Bering Sea. Outside the breeding season, the adult male Pribilof seals spend the winter in the north, but the females and juveniles range widely down the coasts of the northern Pacific, as far south as Japan and

California. When the Pribilof Islands were discovered in 1786, there were over 2 million fur seals there, but the popularity of their soft underfur led to unrestricted sealing and drastic diminution of the herd. International agreement of the North Pacific nations led to research and supervised sealing, and the herds have now built up to their present numbers of nearly 2 million.

A month of commercial sealing is allowed in June, when a regulated number of young males of a specified length are taken. A close watch is also kept on the total numbers of the herd. The blubber is removed from the skins, which are then packed in salt. Many of the skins are processed in the United States, and 3 months and 125 operations separate the raw skin from the finished fur coat. The long rough guard hairs are removed, the soft thick underfur is dyed, and the skin is treated to make it supple. The seal carcasses are converted

into meal for chicken food, and the blubber oil is used in soap.

It is the soft underfur that is commercially desirable. The seal coat is composed of longer, stiffer, guard hairs covering a layer of fine, soft, chestnut-colored underfur hairs. Underfur hairs are relatively sparse in most seals, but abundant in fur seals. The more deeply rooted guard hairs are removed by abrading the inner surface of the skin.

Pribilof seals are dark brown in color with a characteristic short pointed snout. Adult males are about 2.1 meters (82 inches) in length; the females, slightly smaller. Pups are 66 centimeters in length and have a black coat.

Fur seals of the genus *Arctocephalus* are distributed mainly on the shores of Subantarctic islands. They are found on the coasts and offshore islands of Australia, New Zealand, South America, and South Africa, and on more isolated islands

▼ Cape fur seals. This species frequents African and southern Australian coasts. Like sealions, fur seals have visible ears and are able to use their hindlimbs when moving about on land.

such as Macquarie, South Georgia, Bouvet, and Marion. Those on Guadalupe, Galapagos, and Juan Fernandez islands are less known, but a reasonable amount is known about the others. All these fur seals suffered from indiscriminate sealing in the nineteenth century, and the populations are still, even now, building up their numbers. At the present time only the Uruguayan populations of the South American fur seal *A. australis* and the South African fur seal *A. pusillus pusillus* are taken on a carefully controlled commercial basis.

The nose to tail length of adult males of the *Arctocephalus* seals ranges from 1.5 to 2.2 meters (60 to 86 inches), with the Galapagos animals being the smallest and the South African and Australian ones the largest. The adult females are slightly smaller. The newborn pups range from 60 to 80 centimeters (24 to 32 inches) in length.

All these fur seals are very much the same in color: the adult males are a dark blackish brown, with longer hairs on the neck forming a rough mane. Females are brownish gray with a slightly lighter belly. Only the Subantarctic fur seal *A. tropicalis* has a distinctive creamy colored chest and face, and a brush of longer hairs on the head which rise to form a crest when it is agitated. Newborn pups have soft black coats which are

molted for an adult coat at about 3 to 4 months.

Walrus

As the original Pacific walruses of the Oligocene died out, a group invaded the Atlantic 7 to 8 million years ago and developed into the modern animals. The 3 meter (10 foot), 1.2 tonne (2,600 pound) bulky body, the brown wrinkled sparsely haired skin and the long tusks make walruses unmistakable. They live in the shallower waters of the Arctic seas and, being social, haul out in large groups on moving pack ice.

The long tusks, present in both sexes, are the modified upper canines, which grow to 35 centimeters (26 inches) long, though record tusks of 1 meter (40 inches) are known. The ivory has a characteristic granular appearance, easily recognized in carvings. The tusks and strong whiskers are used to disturb the sediment to find mollusks, whose soft parts are sucked out of the shells and eaten.

With a gestation period of fifteen months, a walrus can produce only one calf every two years, and older females reproduce less frequently. The calves, which are 1.2 meters (43 inches) long, suckle and remain with their mother for two years, sheltering under her chest from the worst weather.

▼ A bachelor herd of walruses. The tusks, present in both sexes, are used in seeking out mussels from sediment on the sea floor. Long tusks are also associated with dominance in males.

Leonard Lee Rue/Bruce Coleman Ltd

The pharynx walls in the adult male walrus are very elastic and are expanded as a pair of pouches between the muscles of the neck. These pouches are inflated and used as buoys, and also as a resonance chamber to enhance the bell-like note produced during the breeding season.

Phocids

Phocids evolved round the margins of the North Atlantic from a stock of primitive mustelids (weasels and their relatives), but had become obvious phocids by the mid-Miocene (14 million years ago). The phocids, or true seals, are divided into 'northern phocids', and 'southern phocids', though this subdivision is not precise.

Northern phocids The northern phocids, four genera and ten species of seals, inhabit the temperate and Arctic waters of the Northern Hemisphere. Bearded seals *Erignathus barbatus* and ringed seals *Phoca hispida* live in the high Arctic: the ringed seal is the commonest seal of the Arctic, found up to the North Pole, anywhere there is open water in the fast ice. The bearded seal is so called because of its profusion of moustachial whiskers, which are curious in that they curl into spirals at their tips when dry.

The harp seals *P. groenlandica* and hooded seals *Cystophora cristata* live in the sea around Labrador and Greenland, the harp seal extending to the Arctic coast of Russia. The commercial exploitation of the appealing white-coated pups of the harp seal has incurred universal wrath. Hooded seal pups, known as blue-backs, are also very attractive in their first coats, which are steely blue dorsally and white ventrally. These are also hunted.

The hooded seal gets its name from the enlargement of the nasal cavity of the male. This forms a "cushion" on top of the head, which increases in size with age, and is inflated with air when the nostrils are closed. Hooded seals are also able to blow the very extensible membranous part of the internasal septum out through one nostril, forming a curious red "balloon". The inflation of the hood and balloon occurs both when the seal is excited and in periods of calm.

In the North Pacific, between approximately the Sea of Okhotsk and the Bering Sea, live the ribbon seals *P. fasciata* and larga seals *P. largha*. Harbor seals *P. vitulina* and gray seals *Halichoerus grypus* are found on the coasts on either side of both North Pacific and Atlantic Oceans; Caspian seals *P. caspica* are restricted to the Caspian Sea, and Baikal seals *P. sibirica* to Lake Baikal—the deep lake in eastern Russia.

Most phocids are grayish dorsally and lighter ventrally, plain or with spots, blotches or rings according to species. Only the harp and ribbon seals are strikingly marked. The harp seal has a distinctive black horseshoe mark on its back, while the ribbon seal has a dark chocolate brown coat with wide white bands round the neck, hind end,

Bruce Coleman Ltd

and each fore flipper.

Apart from most of the harbor seals, and the hooded and bearded seals, the northern phocids are born with a first coat of white woolly hair, which is shed after two to three weeks for a coat much like that of the adult. Bearded seals have a gray woolly first coat, and the others shed the first coat before they are born.

Most northern phocids range between 1.3 and 1.8 meters long (52 to 70 inches), and their pups are 65 to 90 centimeters (25 to 36 inches) long. Gray, hooded, and bearded seals are 2.2 to 2.7 meters (86 to 106 inches) long, with hooded and bearded seal pups measuring 1 to 1.3 meters (40 to 52 inches) long; gray seal pups are smaller, at 76 centimeters (29 inches) long.

▲ *Female harp seal and pup. Harp seals are typical phocids, lacking external ears and having the hindlimbs turned backward to form a tail fin. Locomotion on land is by an ungainly "caterpillar" crawl.*

▼ *The nasal cavity of a male hooded seal extends into an inflatable pouch of skin (the "hood") on the top of the head. By closing one nostril and exhaling, the male is able to evert the internasal membrane through the other nostril, forming a red "balloon". The function of this bizarre display is not understood.*

Norman R. Lightfoot/Bruce Coleman Ltd

Frans Lanting/Bruce Coleman Ltd

▲ Phocids such as the monk seal shown here are so streamlined that they have no external ears. Nevertheless, they have excellent hearing. Many species use ultrasonic echolocation to navigate in dark waters and to find food.

▼ The crabeater seal does not eat crabs. When the jaws are closed, its teeth form a sieve that enables it to filter small crustaceans (krill) from water taken into its mouth—like a baleen whale.

Southern phocids Southern phocids are larger than northern seals, measuring 2.2 to 3 meters (86 to 128 inches) long; their pups measure 80 to 160 centimeters (32 to 62 inches). Many skull and skeletal characters also separate the northern and southern phocids, but the six southern genera are not entirely confined to the Southern Hemisphere.

The West Indian monk seal *Monachus tropicalis* is, unfortunately, probably extinct. It lived in the Caribbean, and, though it was seen by Columbus in the fifteenth century, there have been no undoubted sightings for the last thirty years.

The main center of the Mediterranean monk seal *Monachus monachus* is the Aegean Sea, with smaller numbers at suitable spots from Cyprus to Cap Blanc on the Atlantic coast of Mauritania. Early Greek writers such as Homer knew and wrote of this monk seal, but increasing human traffic and pollution in the Mediterranean has disturbed it to such an extent that its numbers are

declining and probably only 1,000 animals are now left. Officially it is protected, and education is alerting the inhabitants of the area to its plight.

On the western atolls of the Hawaiian Islands lives the laysan monk seal *M. schauinslandi*. It is also sensitive to disturbance, and air bases and tourism have reduced its numbers. The population of these monk seals is declining rapidly, and only about 700 animals may be left.

Monk seals are about 2.5 meters (98 inches) long and brown to gray in color, the Mediterranean seal frequently with a white ventral patch.

The Weddell, Ross, crabeater, and leopard seals are the Antarctic phocids. The Weddell seal *Leptonychotes weddelli* is the most southerly, found on fast ice close to land. Although reasonably common, these seals are not gregarious and tend to come singly to breathing holes in the ice. In spite of the hostile environment, these seals have been extensively studied, and it is known that they are particularly deep divers.

The crabeater *Lobodon carcinophagus* is probably the most abundant seal in the world. It is circumpolar and lives in the open seas, and is found on drifting pack ice. Its many-cusped cheek teeth interdigitate and sieve from the water the shrimp-like krill on which it feeds.

Ross seals *Ommatophoca rossi* are found in heavy pack ice, with the greatest numbers in King Haakon VII Sea. Little is known about its life, but it seems to be adapted for rapid swimming and fast maneuvers in order to catch squid and octopus.

The solitary leopard seal *Hydrurga leptonyx* lives in the outer fringes of the pack ice, but immature animals wander far, and they are not infrequent visitors to the shores of New Zealand and Australia. The sinister reputation of this seal is not deserved. Like all carnivores, unless annoyed it is really only aggressive to its potential food, which can be anything from krill to fish, penguins, and even carrion, but its large three-pronged cheek teeth and big gape make an impressive display.

Elephant seals, at 4 to 5 meters (13 to 16 feet) long (pups measure 1.2 meters or 4 feet), the largest of the pinnipeds, are named from the pendulous proboscis that overhangs the mouth so that the nostrils open downwards. It is present only in males and attains its full size by the time the animal is 8 years old. The proboscis is most obvious during the breeding season and, when inflated, acts as a resonator for territorial roaring.

The southern elephant seal *Mirounga leonina* is circumpolar and found on most of the Subantarctic islands; the largest breeding populations are found on South Georgia, Kerguelen and Macquarie Islands. The northern elephant seal *M. angustirostris* occurs mainly on islands off the coast of California.

In the mid-1800s indiscriminate sealing for the oil of these seals nearly exterminated both species. The northern elephant seal has since made a

Francisco Erize/Bruce Coleman Ltd

remarkable recovery and numbers are increasing. The southern elephant seal did not suffer so much, and as its numbers increased, commercial licensed sealing started again. This stopped in 1964 and the seals are now fully protected.

Diving

All seals are able to stay under water for long periods, and some can dive to great depths. Weddell seals may stay under for 73 minutes and dive to 600 meters (2,000 feet), and recent research has found that southern elephant seals can dive to 1,200 meters (4,000 feet) and stay under for nearly 2 hours. Although these long dives are possible, 20 to 30 minutes is the average.

Seals have a lot of blood (12 percent of body weight, compared with 7 percent in humans), which carries a lot of oxygen. During diving, the peripheral arteries are constricted, so most of the blood goes to the brain, and the slowing of the heart rate keeps the blood pressure normal. Phocids can dive deep and long, but otariids, whose blood has a lower oxygen capacity, make shorter, shallower dives. Seals do not suffer from the bends, as they dive with very little air in their lungs and do not have a continuous air supply while diving.

Reproduction

Most seals produce their single pup in spring or summer, but there is considerable variation and one cannot generalize accurately.

Otariids spend much of the year at sea. At the beginning of the breeding season the dominant bulls arrive, select, and defend their chosen territory, and try to stop cows from leaving. The pups are born, head or hind flippers first, shortly after the females arrive, and about one to two weeks later, depending on species, the females mate again. There is a two to five month delay before the blastocyst becomes implanted and active gestation starts. The total gestation period is, therefore, about 11½ months, and the active gestation period, 7 to 8 months, depending on species. The pups start to suckle very shortly after birth, and most otariid pups are suckled for about a year and during this time become increasingly adventurous and start to find some solid food. The mother stays with her pup, on land, for about two weeks, and then goes to sea to feed, returning at intervals to suckle her pup. Pups frequently gather in groups, or pods, while their mothers are at sea, and mother–pup recognition is by call. Bulls ignore the pups, but will remain on land guarding their territory until the end of the breeding season, going without food for about two months, but details vary according to species. The Antarctic fur seal and the northern fur seal suckle for the shorter time of three to four months, and the walrus for two years. After a few months, the harem system breaks down and the animals disperse.

Dr Eckart Pott/Bruce Coleman Ltd

Amongst the phocids the territory system is found in gray and elephant seals, but in the others there may be concentrations of animals at breeding time, or just pairs coming together. Mating takes place from twelve days to seven weeks after birth, and the length of lactation is equally variable (twelve days to two months), depending on species. There is a delay of two to five months before the blastocyst is implanted. After lactation the seals disperse.

Seal milk is particularly rich in fat and protein: nearly 50 percent is fat (compared with 3.5 percent in cow's milk), and this is correlated with the fast rate of growth of the pup. Phocid pups have a high daily increase in weight, ranging from 1.3 kilograms (2¾ pounds) in gray seals, 2.5 kilograms (5½ pounds) in harp seals to 6 kilograms (13¾ pounds) in southern elephant seals. With the longer lactation period in otariids, the daily weight gain is lower (about 40 to 60 grams or 1½ to 2 ounces), though the two fur seals with shorter lactation periods (Antarctic and northern fur seals) have pups that put on weight faster (about 100 grams or 4 ounces a day). It is also necessary for all seal pups to lay down a layer of insulating blubber as soon as possible. There is virtually no lactose in the milk and as seals do not tolerate it, orphan pups must be given a specially designed formula, not cow's milk.

JUDITH E. KING

▲ A herd of northern elephant seals, consisting of a male with his harem and their offspring. Reaching a weight of 2.4 tonnes (5,300 pounds), male elephant seals are the largest of the pinnipeds and indeed the largest carnivores. The southern elephant seal has the amazing capacity to remain below the surface for nearly two hours, while reaching a depth of as much as 1,200 metres (3/4 mile).

WHALES & DOLPHINS

Heaviside's dolphin
Cephalorhynchus heavisidii
Head–tail length: 1.4 m (4.5 ft)
Weight: 50 kg (110 lb)

Blue whale *Balaenoptera musculus*
Head–body length: up to 30 m
(100 ft)
Weight: up to 130 tonnes
(286,500 lb)

CONSERVATION WATCH
!!! The vaquita *Phocoena sinus*
and baiji *Lipotes vexillifer* are
listed as critically endangered.
!! The endangered species are:
northern right whale *Eubalaena
glacialis*; Sei whale *Balaenoptera
borealis*; blue whale *Balaenoptera
musculus*; fin whale *Balaenoptera
physalus*; bowhead whale *Balaena
mysticetus*, finless porpoise
Neophocaena phocaenoides, sperm
whale *Physeter catodon*; Ganges
river dolphin *Platanista
gangetica*; Indus river dolphin
Platanista minor.
! 6 species are vulnerable.

▼ *River dolphins have small,
degenerate eyes. Because they usually
live in turbid water, they navigate and
find their prey by echolocation. The large
"melon" on the forehead acts as a lens
for ultrasonic vibrations emitted from the
nasal cavity.*

Whales and dolphins are found in all seas of the world, and in some rivers and lakes. Their streamlined bodies and their ability to dive for long periods and to great depths are just two of their many adaptations to life under water. The order Cetacea contains two extant suborders, the whalebone or baleen whales (Mysticeti), and the toothed whales (Odontoceti). A third suborder, the Archaeoceti, contains only extinct forms.

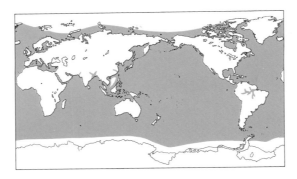

TOOTHED WHALES

There are six families of toothed whales. Many species are confined to relatively small areas of ocean, although the different species are found in a variety of habitats. Among the dolphins, some species are found only in rivers, some only in onshore waters, and yet others only in the open ocean. On the other hand some species, such as the killer whale and the bottlenose dolphin, are seen in both inshore and deep ocean waters.

Sperm whales

The family of sperm whales, Physeteridae, includes the largest and in some ways the most specialized of the toothed whales. Sperm whales have a very large head that contains a structure known as the spermaceti organ, and functional teeth only in the lower jaw, which is underslung.

White whales

The white whales, Monodontidae, have no beak; the dorsal fin is absent, rudimentary, or small; there are no grooves on the throat; and the neck vertebrae are all separate, providing some flexibility to the neck. Males of one member of the family, the narwhal, have a highly developed tooth in the upper jaw that forms a long, spiral tusk.

Beaked whales

The beaked whales, Ziphiidae, derive their common name from the characteristic long and protruding snout. A pair of throat grooves is present, and there is no notch in the middle of the trailing edge of the flukes. In all but one species there are two teeth in the lower jaw, but they protrude through the gums only in adult males.

Dolphins

Although often referred to as the "true dolphins", the family Delphinidae includes species that are not distinctly "dolphin-like", such as killer whales and pilot whales. Delphinids, the most abundant and varied of all cetaceans, are found from tropical to polar seas. Many of the smaller species have a beak, or elongation of the upper and lower jaws.

Porpoises

Porpoises, the Phocoenidae, are small toothed whales with the following characteristics: body length of less than 2.45 meters (8 feet); more than five neck vertebrae fused; and more than fifteen teeth in the back row of the upper jaw. None of the porpoises has a beak.

River dolphins

River dolphins, the Platanistidae, are characterized by an extremely long beak, and a long, low dorsal fin. The family includes four species that inhabit fresh-water rivers in Asia and South America, and one South American coastal species.

Norbert Wu/Planet Earth Pictures

BALEEN WHALES

The three families of baleen, or whalebone, whales are found in all oceans, and most undertake extensive migrations which take them across and between oceans.

Right whales

Right whales, the Balaenidae, are characterized by long and narrow baleen plates, known as whalebone, and a highly arched upper jaw, which distinguish them from gray whales and rorquals. Other features include a disproportionately large head (more than a quarter of the total body length), a long thin snout, huge lower lips, and fused neck vertebrae. The skin over the throat is smooth, and there is no dorsal fin.

Gray whales

Gray whales, the Eschrichtiidae, have no throat grooves, although two to four furrows are present on the throat. There is no dorsal fin, but there are several humps on the upper surface of the tail stock. The neck vertebrae are not fused together.

Rorquals

Rorquals, the Balaenopteridae; are baleen whales that have relatively short, triangular baleen plates. The head length is less than a quarter of the body length; numerous grooves on the throat extend from the chin to the middle abdomen; and there is a dorsal fin, which is often small. The upper jaw is relatively long; the lower jaw bows outwards; and the neck vertebrae usually are not fused.

▼ *Killer whales cooperate in herding prey into shallow water; as well as eating large fishes, they prey upon warm-blooded animals and may patrol shores in search of unwary seals or penguins.*

Francois Gohier/Ardea London

▲ *Weighing up to 50 tonnes (110,000 pounds), the humpback whale is one of the largest filter-feeders. It eats small schooling fishes and krill—7 centimeter (3 inch) crustaceans that occur in dense populations in polar seas, particularly the Antarctic Ocean.*

TORPEDO-SHAPED FOR LIFE IN THE SEA

The torpedo-shaped bodies of cetaceans are very efficient for moving through water. The forelimbs are rigid paddles that help in balancing and steering. Most, but not all, species have a dorsal fin. There are no external ears.

The skin is smooth and virtually hairless, although fetuses bear a few hair follicles on the snout which in certain species persist throughout life. Sweat and sebaceous glands are lacking.

Immediately beneath the skin is an insulating layer of blubber, which contains much fibrous tissue interspersed with fat and oil. The color of whales varies from all white to all black, with varying shades of gray in between. Many species are dark gray or black along the back and top of the head and flukes, and light gray or white beneath. Some bear stripes, spots, or patches of black, gray, white, or brown. The most striking of these are the killer whale, with its distinctive black and white markings, the spotted and striped dolphins, and the common dolphin, which has an exquisite coloring of dark gray or black above, light beneath,

and patches of gray and ochre in a figure-of-eight pattern along each flank. Toothed whales have numerous simple, conical teeth and telescoped, asymmetrical skulls. Baleen whales have no visible teeth but plates of baleen suspended from the hard palate, and symmetrical skulls. Baleen or whalebone is a horny substance that was formerly used for stiffening corsets. The real bones are spongy in texture and filled with oil. There is little or no visible neck, and the seven neck vertebrae are foreshortened. Some of them are fused in some species. There are no skeletal supports for the tail flukes or for the dorsal fin (if there is one).

The pelvic girdle is represented by two small rods of bone embedded in the muscles of the abdominal wall. They do not articulate with the backbone. In some species a projection from the pelvic bone represents the vestigal thigh bone.

Whales can attain great size because they live entirely in water and need not support their own weight. The blue whale, which grows to almost 30 meters (33 yards) in length, is the largest animal that has ever inhabited the earth.

Howard Hall/Planet Earth Pictures

◄ Dolphins swim very fast by relatively small movements of the horizontal tail fin. Their speed is facilitated by the streamlined body and by a layer of subcutaneous fat (blubber), which permits the skin to move in such a way that eddies (which cause friction) are minimized.

HOW WHALES BREATHE

Whales breathe less frequently than land mammals, and they can hold their breath for extraordinarily long periods during dives. Although their lung capacity is no greater than that of land mammals of equivalent size, whales take deeper breaths and extract more oxygen from the air they breathe, and exchange more air in the lungs with each breath. A whale's lungs are at least partially inflated when it dives, unlike the seal, which exhales before diving.

The nostrils are modified to form a blowhole at the top of the head. It is single in the toothed whales, double in the baleen whales. The skin immediately surrounding the blowhole has many specialized nerve endings, which are very sensitive to the change as the blowhole breaks the water surface. Breathing out and in again often occurs extremely rapidly, in the fraction of a second the blowhole is above the surface. The blowhole is closed when the animal is submerged by the action of muscles on a series of valves and plugs.

When a whale surfaces and exhales, a spout or "blow" is seen. This is composed of moist air exhaled under pressure, together with some condensation of water vapor as the pressure drops on release from the blowhole to the atmosphere. It probably also contains an emulsion of fine oil droplets from cells lining the nasal sinuses, and mucus from the respiratory passages.

► A blue whale "blowing". Baleen whales have two nostrils; toothed whales have a single blowhole.

Francois Gohier/Ardea London

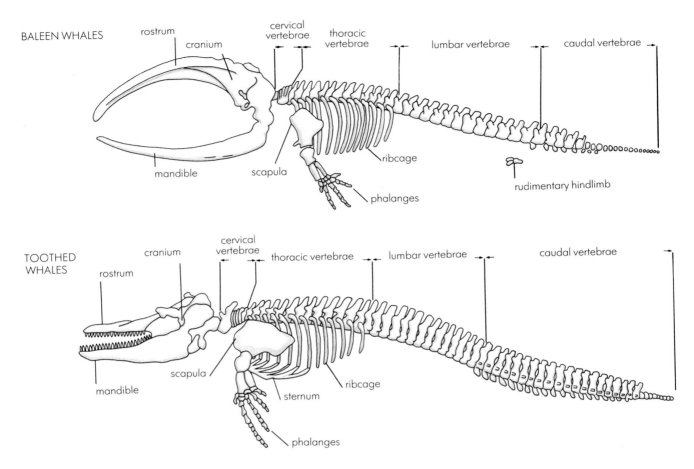

BALEEN WHALES — rostrum, cranium, cervical vertebrae, thoracic vertebrae, lumbar vertebrae, caudal vertebrae, mandible, scapula, phalanges, ribcage, rudimentary hindlimb

TOOTHED WHALES — rostrum, cranium, cervical vertebrae, thoracic vertebrae, lumbar vertebrae, caudal vertebrae, mandible, scapula, sternum, ribcage, phalanges

▲ *There are other differences between baleen and toothed whales besides the presence or absence of teeth, as these drawings of their skeletons show. The skull of a baleen whale is symmetrical; that of a toothed whale is asymmetrical. This is possibly related to the fact that toothed whales have a single blowhole. Sternal ribs are absent from the baleen whales but present in the toothed whales. But the most obvious external difference is size: few species of toothed whales approach the length of the smallest baleen whales.*

KEEPING WARM IN COLD SEAS

The large size of all whales, particularly sperm whales and most baleen whales, gives them significant metabolic advantages, especially in the cold seas where most of them spend part of their life. Blubber is a very effective insulator. It enables significant heat conservation when blood flow to the skin is reduced. The blubber varies in thickness seasonally within species, and there is great variation in its thickness among species. It has been recorded up to 50 centimeters (20 inches) thick in bowhead whales, whereas in bottlenose dolphins it is more in the region of 2 centimeters (1 inch).

When a whale is active, such as when diving and chasing prey, considerable metabolic heat is produced. Excess heat can be dissipated in the surrounding cold water by opening large arteries that pass through the blubber, thereby increasing the amount of warm blood flowing through the skin. Within the flippers, flukes, and dorsal fin is an elaborate arrangement of arteries and veins that provides fine control of heat loss or conservation in response to excess or deficiency.

ECOLOGY AND BEHAVIOR

The social and sexual behavior of whales is widely divergent among species. Most, but not all, baleen whales mate and breed in tropical seas in winter, and migrate to summer feeding areas in high latitudes. Sperm whales form nursery schools

consisting of adult females, calves, and juveniles. As males mature they migrate annually to higher latitudes, and the largest males return and compete for access to the nursery schools during the breeding season.

Spinner dolphins rest during the day in groups of about 20 in inshore bays, moving offshore at night, where they unite to form groups of up to several hundred, and feed at great depth. Killer whales occur in all oceans and exhibit considerable variation in their behavior. They may hunt singly or cooperatively in groups. They tend to live in groups (or pods), which may join to form larger pods, either temporarily or permanently.

SWIMMING AND DIVING

The streamlined shape and smooth skin of cetaceans allow water to flow easily over the body surface. This, as well as the mode of locomotion and subtle changes in conformation of the skin, permit them to move through the water very efficiently. Streamlined flow over the body surface during movement means there is very little friction or drag, enabling some species to achieve great speed using relatively little energy.

Cetaceans propel themselves by up and down movements of the tail, in such a way that the flukes present an inclined surface to the water at all times. The force generated at right angles to the surface of the flukes is resolvable into two components, one

Francisco Erize/Bruce Coleman Limited

raising and lowering the body and the other driving the animal forward. The flippers are used for balancing and steering.

All species can dive to greater depths and remain submerged for longer periods than land mammals. Some species have exceptional diving abilities: for example sperm whales and certain beaked whales are known to dive to more than 1,000 meters (550 fathoms) and stay submerged for more than two hours. Baleen whales and many dolphin and porpoise species, however, generally make shallow dives of less than 100 meters (55 fathoms) and surface after ten minutes or less.

Whales' diving ability is due to anatomical and physiological modifications in the circulatory system, and chemical adaptations in the blood, muscles, and other tissues. The blood makes up a greater proportion of body mass and is capable of carrying significantly more oxygen than that of land mammals. The muscles contain large amounts of myoglobin, which permits increased oxygen storage and gives the muscles a characteristic dark color. When a whale dives, its heart rate slows down. Changes in the distribution of blood in the body mean that those tissues and organs that require oxygen most receive most of the blood. The whale is able to tolerate higher levels of carbon dioxide and other byproducts of metabolism than non-diving mammalian species, which means they can go for considerably longer between breaths

while still remaining active.

Whales have unusual blood vessels, known as retia mirabilia, in the upper part of the chest wall and extending along the vertebrae in the neck and towards the flippers. Although they are very extensive networks of small arteries, and very obvious on postmortem examination, their function is not known. They are present, particularly around the base of the brain, in some land mammals, but they are infinitely more extensive in whales. It has been assumed that they play some role in diving in whales, but precisely what is unknown.

KILLERS AND SIEVERS

The only toothed whale that regularly feeds on animals other than fish and cephalopods such as squid and octopus is the killer whale. It also eats the flesh of warm-blooded animals such as seals, sea lions, dolphins, and penguins. The false killer whale *Pseudorca crassidens* also eats warm-blooded animals occasionally.

The conical teeth of toothed whales are used to seize and hold prey but are not adapted for chewing. The teeth vary in size and number, and in some species, such as the sperm whales and Risso's dolphin *Grampus griseus*, they are absent from the upper jaw. Whereas Risso's dolphin has an average of only six to eight teeth in the lower jaw, the common dolphin has more than a hundred in each

▲ The bottlenose dolphin is familiar to millions of visitors to marine mammal shows. Under natural conditions it leaps out of the water and, with training, it will do so on demand. Although dolphins have large brains, they are no more intelligent than many other mammals.

▲ *Humpback whales feeding near the surface. They swim below a school of fishes and confuse them by emitting a multitude of small bubbles. They then swim upward with the mouth open, closing it as they break the surface. Water is squeezed out of the mouth, through the sieve of baleen plates, by raising the tongue. The concentrated mass of fish is then swallowed.*

jaw. The teeth have not come through the gums at birth; they emerge at several weeks or months of age, and those teeth remain throughout life. A thin lengthwise section of a whale's tooth reveals layers, which provide an accurate index of age.

The baleen of whalebone whales occurs as horny plates arranged along either side of the upper jaw, more or less like the leaves of a book, with the inner portion frayed into bristle-like fibers. They form a mesh that sieves the very small animals, collectively called plankton, that make up the diet of baleen whales. The composition of the plankton depends on the region in which whales feed, and to some extent different species feed on different kinds of plankton.

SIGHT, SOUND, AND TOUCH

Vision is considered to be quite acute in whales, although it is uncertain what its relative importance is. The eye is adapted to permit quite good vision both under water and in air, and at depths where light intensity is very low.

Whales produce a wide range of underwater sounds, some of which almost certainly are used for communication between individuals. The long and relatively complex song of humpback whales has attracted considerable interest, although its significance is not fully understood. It differs with location, and also within areas over time.

Toothed whales produce audible clicks, which are used in echolocation. They probably depend to a large extent on echolocation for orientation and to obtain food. It has been suggested that some toothed whales may stun prey with accurately

directed bursts of very high energy sound.

Whales have no external ear, but the middle ear receives sound through the tissues of the head and transmits it to the inner ear. The two ears are insulated by an enveloping layer of albuminous foam, allowing whales a three-dimensional perception of the direction and distance from which sound is coming. The ear canal of baleen whales contains a waxy plug, which in section shows layers that are used to determine the age of individual animals, just as tooth sections are used in toothed whales. Whales have no organ for smelling and therefore no sense of smell. They have some modified taste buds in the mouth, but their function is not known.

Whales have a highly developed sense of touch, which involves the entire body surface. Particular areas of the skin have specialized functions: for example, in toothed whales the skin in the region of the jaw may be used to "feel" and to detect sources of low-frequency vibrations.

REPRODUCTION AND DEVELOPMENT

Reproductive behavior is known to vary among species, although little is known about it in most. Breeding is seasonal in migrating species, but in other species it seems to occur through most, if not all, the year. The release of each egg leaves a permanent scar on the ovary of all species. This provides a permanent record of past ovulations, and has been used to estimate age in some large whales. Males of some species have a breeding season, while in others sperm are present all year.

The gestation period in most whale species is 11

to 16 months. A single young is born; twins are very rare. The newborn young is approximately one-third the length of the mother. (Body length is measured as the straight-line length from the tip of the upper jaw to the depth of the notch between the tail flukes.) The teats of the mammary glands lie within small paired slits one on either side of the genital opening. The mammary glands lie beneath the blubber over an extensive area of the abdominal wall. Milk is forced into the calf's mouth during suckling by contraction of muscle overlying the mammary glands. Sucking calves grow very rapidly, a fact that is associated with the high proportions of fat and protein in the milk.

CONSERVATION

For as long as humans have been on the Earth, whales have been considered an exploitable resource. We have killed many species of whales, particularly the large whales, for products such as oil, meat, baleen, and hormones.

As recently as the 1960s we became generally aware of the need to conserve the Earth's resources. The proposition that human use of nature should be placed on a rational basis became widely accepted, and we recognized the need to conserve species of whales. Gradually since then, measures have been introduced to try to achieve this end.

The measures taken have varied with the species and with the cause of the population decline. With a few exceptions, the size of whale populations is very difficult to assess. Some species, however, had obviously suffered drastic population declines through over-harvesting, and this problem was the first to be addressed. Drastic action was taken in some cases: for example the total ban on killing southern humpback whales introduced in 1963. Quotas for killing whales, often rather complex and, in some cases, of limited effectiveness for the most endangered species, were introduced. In very

recent times, whale harvesting has been reduced to a low level, and only those nations supporting so-called scientific whaling are involved.

New conservation problems have now arisen, however, affecting in particular some oceanic species of small toothed whales and dolphins. Many thousands of dolphins have been drowned in the huge gillnets, or driftnets, that are set to catch fish in the seas. Purse-seine nets for capturing tuna have resulted in the deaths of millions of small toothed whales, particularly spotted and spinner dolphins and, less frequently, common dolphins. The annual mortality of dolphins caused by these industries is monitored, and attempts have been made to modify the netting procedures. But the world's most seriously endangered cetaceans include some of the river dolphins, notably the Chinese river dolphin that inhabits the Yangtze River. Netting, river pollution, and increased boat traffic have led to what could prove to be intolerable pressures on these animals.

M. M. BRYDEN

▲ A minke whale being butchered in Iceland. Organized deep-sea whaling, using harpoon guns, began in the 1860s. A century later there were so few large whales left that some species were on the verge of extinction. Most nations now ban whaling.

Kenneth W. Fink/Ardea London

WHY DO WHALES RUN AGROUND?

Like many other animals, from bacteria to mammals, whales appear to have a sensory faculty that can receive directional information from the Earth's magnetic field. Magnetite crystals have been observed in the tissues surrounding the brain of several species of toothed whales.

The total magnetic field of the Earth is not uniform, but is locally distorted by the magnetic characteristics of the underlying geology. In the floors of the oceans, the movements of the continents have produced series of almost parallel magnetic "contours", which could be used by whales in navigation. Evidence is mounting that whales do use this sense in navigation, and that the phenomenon of mass strandings of live whales results from navigational mistakes.

These sperm whales, disoriented in shallow water, were stranded at low tide off the coast of south-eastern Australia.

John R. Brownlie/Bruce Coleman Limited

SEA COWS

ORDER SIRENIA
• 2 families • 2 genera • 4 species

SIZE

Dugong *Dugong dugon*
Head–tail length: up to 3.3 m
(11 ft)
Weight: up to 500 kg (1,100 lb)

Manatees (genus *Trichechus*)
Head–tail length: up to 3.9 m
(13 ft)
Weight: up to 1,500 kg (3,300 lb)

Steller's sea cow *Hydrodamalis
gigas* (now extinct)
Head–tail length: 8 m (26 ft)
Weight: up to 5.9 tonnes
(13,000 lb)

CONSERVATION WATCH
! The 4 species in this order—
dugong *Dugong dugon;*
Amazonian manatee *Trichechus
inunguis;* Caribbean manatee
Trichechus manatus; African
manatee *Trichechus senegalensis*—
are all listed as vulnerable.

The forerunners of the modern sirenians, or sea cows, are believed to have been descended from an ancestor shared with elephants. Even though they have persisted since the Eocene epoch (57 million to 37 million years ago), there have never been many different types of sea cows. There are far fewer species of seagrasses, on which the sea cows feed, than terrestrial grasses, and this is considered to be a major reason for the lack of diversity among sirenians. Only four species of sea cow survive today: one dugong (family Dugongidae) and three manatees (family Trichechidae). All are grouped in the obscure order Sirenia. A fifth species, Steller's sea cow, was exterminated by humans in the eighteenth century. Skeletal remains of this giant kelp-eating dugong, the only cold-water sirenian, can be seen in many major museums.

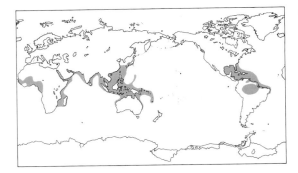

UNDERWATER GRASS EATERS

Manatees appeared during the Miocene (24 million to 5 million years ago), when conditions favored the growth of fresh-water plants in the rivers along the South American coast. Unlike seagrasses, these plants had high concentrations of silica, which rapidly wears away teeth. The manatees countered this problem by developing a system of continuous replacement of their molar teeth: throughout life worn teeth are shed at the front of the jaw and replaced at the back. Dugongs have a different method of coping with tooth wear: their two rear molars are open-rooted and continue to grow. However, the crowns of these teeth are worn continuously, which means that they are very simple and peg-like. In addition, the angle of the dugong's snout is more pronounced than that of the manatee, so the dugong is virtually obliged to feed on bottom-dwelling organisms, whereas the manatee can also feed on plants growing at or near the water's surface. These differences seem to have given manatees an edge over dugongs, which they have apparently displaced from the Atlantic area.

The four extant species of sea cows inhabit largely separate geographical ranges and are restricted to the tropics and subtropics. The dugong's range spans the coastal and island waters of 43 countries in the Indo-Pacific region between East Africa and Vanuatu. In contrast to dugongs, which are restricted to the sea, manatees mainly occur in rivers and estuaries. The West Indian and West African manatees occupy similar

fresh-water and marine habitats on either side of the Atlantic. Their supposed common ancestor is believed to have migrated to Africa across the Atlantic. Amazonian manatees are restricted to fresh-water habitats in the Amazon basin.

As aquatic herbivores, sirenians have characteristics of both marine mammals and terrestrial herbivores. The dugong's body is shaped like a dolphin's but is a little less streamlined, and the tail flukes are like those of a whale or dolphin. The external ears also resemble those of whales and dolphins in being merely small holes, one on each side of the head. The dugong's hearing is acute. The nostrils close with valve-like flaps and are on top of the head so the dugong can surface without its body breaking the water. But the dugong's head, with its vast upper lip covered with stout sensory bristles, is more like that of a pig, and inside the downturned mouth are horny pads like those a cow uses for grasping grass.

Manatees look very like dugongs. The most obvious difference is in the shape of the tail; manatees have a horizontal, paddle-shaped tail rather like that of a beaver or a platypus.

Sirenians are large animals. West Indian manatees average over 3 meters (10 feet) long and weigh over 500 kilograms (1,100 pounds). West African manatees are similar in size, while Amazonian

▼ A female dugong is at least 10 years old before she bears her first calf, which is suckled for up to 18 months. Subsequent births are 3 to 5 years apart.

Ben Cropp/Auscape International

▲ *Manatees range from the sea to rivers. Their diet includes seagrasses, and vegetation floating on the water surface or overhanging the water's edge. The molar teeth continually move forward in the jaws, the oldest being shed as new teeth erupt at the back of the tooth-rows.*

Jeff Foott/Bruce Coleman Limited

manatees and dugongs average about 2.7 meters (9 feet) long and weigh an average of about 300 kilograms (660 pounds). Apart from their size, sea cows have few defenses. Mature male dugongs have tusks, which they apparently use in the fighting that precedes mating. Fortunately, dugongs and manatees have few natural predators.

ATTENTIVE MOTHERS

The life-span of sirenians is long and their reproductive rate low. The age of dugongs can be worked out by counting the layers laid down each year in their tusks, like growth rings in a tree. Individuals may live for seventy years or more, but a female does not have her first calf until she is at least ten years old, and then only bears a single calf every three to five years, after a gestation period of about a year. Information from manatees whose ages are known indicates that they may have a slightly higher reproductive rate, bearing their first calves at a minimum age of three years and having an average calving interval of two to five years.

Female dugongs and manatees are very attentive mothers, communicating with their calves by means of bird-like chirps and high-pitched squeaks and squeals. A young calf never ventures far from its mother and frequently rides on her back. Although it starts eating plants soon after birth, a calf continues to nurse from mammary glands near the base of its mother's flippers until it is up to 18 months old. The mammae look rather like the breasts of a human female, a probable reason for the belief that dugong or manatees form the basis of the mermaid legend.

CONSERVATION

Dugongs and manatees have delicious meat, docile natures, few defenses, low reproductive rates, and ranges that span the waters of many protein-starved developing countries. It is no wonder that all are classified as vulnerable to extinction by the International Union for the Conservation of Nature.

If you ask Aborigines from the coastal regions of northern Australia to name their favorite food, they are likely to choose dugong. This choice is endorsed by coastal peoples throughout much of the tropical and subtropical regions of the world who prize the meat of sea cows. The meat does not taste of fish, but has been variously likened to veal, beef, or pork because, like the domestic land mammals we use for meat, sea cows feed primarily on plants.

Although successfully exploited by indigenous peoples for thousands of years, dugongs and manatees were not able to sustain the increased exploitation that often followed European colonization. Steller's sea cow was extinct within 30 years of its discovery by a Russian exploration party in the Commander Islands off the coast of Alaska in the eighteenth century. Today, dugongs and manatees are protected throughout most of their ranges, except in some areas where traditional hunting is permitted. But this protection is difficult to enforce, particularly in developing countries that cannot afford wildlife wardens. Even if it were possible to prevent all deliberate killings of sirenians, their preservation would not be assured. Dugongs and manatees often accidentally drown in gill nets set by commercial fishermen. In Florida, most manatees have distinctive scars inflicted by boat propellers, and collisions with boats constitute the largest identifiable source of mortality.

The most serious threat to sirenians, however, comes from the loss and degradation of their habitat. This is most obvious in Florida, home to the only significant populations of the West Indian manatee, where there has been rapid growth in the human population. Attempts to reduce manatee mortality in this region have had limited success.

The prognosis for the dugong is brighter, particularly in Australia, where human population pressure is low throughout its range, except in southeast Queensland. Dugong population estimates based on aerial surveys in Australia add up to more than 80,000 and not all the suitable habitat has been surveyed. Dugong sanctuaries have been established in some important dugong habitats, particularly in the Great Barrier Reef region. However, because dugongs have such a low reproductive rate and because of the difficulties of censusing them, it will probably be a decade before it can be determined whether these conservation initiatives are working.

In the last two decades we have learned a lot about dugongs and manatees. It remains to be seen if this knowledge can be applied to ensure their continued survival. The most effective action would be to identify and protect the habitats that still support significant numbers of these "gentle outliers in the spectrum of mammalian evolution".

HELENE MARSH

ELEPHANTS

Elephants are classified in the order Proboscidea, after the most distinguishing feature of these mammals—the proboscis or trunk. Ancestors of the modern elephants are believed to have lived 55 million years ago; they occupied extreme environments, from desert to tropical rainforest and from sea level to high altitudes. With the exception of Antarctica, Australia, and some oceanic islands, the proboscideans have at some time inhabited every continent on this planet.

Asian elephant *Elephas maximus*
Head–body length: 300–640 cm (120–250 in)
Shoulder height: 200–335 cm (80–135 in)
Weight: to 5.4 tonnes (12,000 lb)

African elephant *Loxodonta africana*
Head–body length: 375–550 cm (148–217 in)
Shoulder height: 240–400 cm (95–160 in)
Weight: up to 7 tonnes (15,400 lb)

CONSERVATION WATCH
!! Both species are endangered.

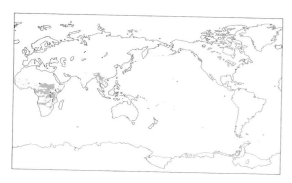

AN IMPORTANT ECOLOGICAL ROLE

According to some authors, 352 proboscidean species and subspecies have been identified; of these 350 are extinct. Parts or complete carcasses of the extinct woolly mammoth *Mammuthus*

primigenius have been discovered in the permafrost of the Arctic Circle. One of the best known is that of 'Dima', a 44,000-year-old mammoth calf whose intact frozen body, discovered in June 1977 in the Magadan region of Siberia, provided scientists with a unique opportunity to study both its anatomy and its soft tissues.

The living species belong to two genera, with one species each: the African elephant *Loxodonta africana* and the Asian elephant *Elephas maximus*. Elephants play a pivotal role in their ecosystem. Attributes that make the elephant an inseparable part of their environment include seed dispersal (for example, of acacia trees) through their fecal material, which promotes rapid germination; distribution of nutrients in their dung, which is carried below the ground by termites and dung

▼ *African elephants wallowing. The layer of mud that remains on the skin protects it from insects and sunburn. Elephants that wallow in clean water toss dust onto their backs to provide similar protection.*

Anthony Bannister/NHPA

167

The number of teeth is identical for the two species within the lifetime of an individual. The total includes two upper incisors (tusks), no canines, twelve deciduous premolars, and twelve molars. Unlike most other mammals, elephants do not replace their cheek teeth, molars and premolars, vertically—that is, a new tooth developing and replacing the old one from above or below—but rather in a horizontal progression. A newborn elephant has two or three cheek teeth in one jaw quadrant and, as it ages, new teeth develop from behind and slowly move forward. At the same time the previous teeth are worn away, move forward, and fragment. And so, as in a conveyor belt system, new and bigger teeth replace the old.

SKIN, EARS, AND BRAIN

The old name of the order Proboscidea, "Pachydermata", refers to their thick skin. The thickness varies from paper thin on the inside of the ears and around the mouth and anus, to about 25 millimeters (1 inch) around the back and some areas of the head. Despite its thickness, the skin is a sensitive organ. The bodies of both species of elephant are usually gray and covered with sparse hair; those of African elephants often seem brown or even reddish because the animals wallow in mudholes or plaster colored soil to their skin. Wallowing seems to be an important behavior in elephant societies; the mud, it appears, protects against ultraviolet radiation, insect bites, and moisture loss. Scratching against trees and bathing seem to be equally important behaviors in skin care.

Body heat can be lost through the ears: within a closely related group of mammals, the size of the ears is often related to climate (larger ears in hotter environments) and to size (proportionately larger ears in bigger animals). The African elephant, which is about 20 percent heavier than the Asian species, has ears of more than twice the surface area. That elephants' large ears function as cooling devices can be demonstrated by the large number of blood vessels at the medial side of the ears, where the skin is about 1 to 2 millimeters (1/16 inch) thick, and by the high frequency of ear flapping during warm to hot days when there is little or no wind. Large ears also trap more sound waves than smaller ones.

The brain of an adult elephant weighs 4.5 to 5.5 kilograms (10 to 12 pounds). It possesses highly convoluted cerebrum and cerebellum (in mammals, these control motor and muscle coordination, respectively). The temporal lobes, known to function as memory centers in humans, are relatively large in elephants. The digestive system is, for mammals, simple. The combined length of the small and large intestines may reach 35 meters (about 100 feet), and it takes about 24 hours to digest a meal. Elephants digest only about 44 percent of their food intake.

D. Parer & E. Parer-Cook/Auscape International

▲ *Young African elephant. In addition to their function in hearing, an elephant's ears are used as signals of emotional state and as radiators of excess body heat. African elephants have much larger ears than their Asian relatives.*

beetles (soil aeration); providing water by digging waterholes which other species also use; trapping rainfall in the depressions from their footprints and bodies; enlarging existing waterholes as they plaster mud on their bodies when bathing and wallowing; making paths, usually leading to waterholes, which act as firebreaks; providing food for birds when walking in high grass by disturbing insects and small reptiles or amphibians; and providing protection for other species as they are tall and can see a long way and alert smaller animals to approaching predators.

THE SKELETON

As in all mammalian species, the skeleton of an elephant is divided into four major parts: the skull, vertebral column, appendages, and ribs and sternum. A side view of a skeleton shows the legs are in almost a vertical position under the body, like the legs of a table. This arrangement provides a strong support for the vertebral column, thoracic and abdominal contents, and the great weight of the animal. In other mammals — for instance, a dog or a cat — the legs are at an angle to the body.

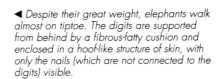

◀ *Despite their great weight, elephants walk almost on tiptoe. The digits are supported from behind by a fibrous-fatty cushion and enclosed in a hoof-like structure of skin, with only the nails (which are not connected to the digits) visible.*

REPRODUCTION

The male reproductive system consists of typical parts found in other mammals. Unlike many mammals, however, the testes in elephants are permanently located inside the body, near the kidneys. The penis, in a fully grown male, is long, muscular, and controlled by voluntary muscles; it can reach 100 centimeters (40 inches) and has a diameter of 16 centimeters (6 inches) at its base. When fully erected, the penis has an S-shape with the tip pointing upwards.

The female reproductive system is also that typical for most mammals, but the cervix is not a distinct structure. The clitoris is a well-developed organ which can be 40 centimeters (15 inches) long, and the vulval opening is located between the female's hindlegs, not under the tail as, for example, in bovids and equines. The mammary glands are located between the forelegs, which enables the mother to be in touch with her calf while it suckles.

Elephants attain sexual maturity between the ages of 8 and 13. The estrous cycle may vary between 12 and 16 weeks, and estrus (ovulation and receptivity) lasts a few days; the egg is viable only about 12 hours. Mating is not confined to any season and the gestation period may last 18 to 22 months — the longest pregnancy of any known living mammal. Copulation may take place at any time in a 24-hour cycle; it is usually performed on land, but elephants have been observed to mate in water. The bull pursues the cow until she is ready to mate, at which time a short interaction period of trunk and body contact may take place, followed by the male elephant mounting the female in the usual quadruped position. The long penis of an adult bull plays an important role in the fertilization process; a young bull with a short penis may be able to mate with a cow in estrus, but it is unlikely that he would impregnate her. The entire copulating process lasts about 60 seconds, at the end of which the two elephants separate and may remain near each other for a short period.

Elephants usually bear a single young. Newborn calves weigh 77 to 136 kilograms (170 to 300 pounds) and are 91 centimeters (3 feet) tall at the shoulder. Young are hairy compared with adult animals; the amount of hair reduces with age. Calves may consume 11.4 liters (3 gallons) of milk a day. Weaning is a very gradual process, which begins during the first year of life and may continue until the seventh or sometimes the tenth

THE ELEPHANT'S TRUNK

The trunk of an elephant is its single most important feature; it is an indispensable tool in everyday life. Anatomically, it is a union of the nose and the upper lip. Early naturalists described the trunk as "the elephant's hand" or as "the snake hand". Its flexibility and maneuverability are certainly extraordinary. It is said that an elephant can pick up a needle from the ground and bring it to its trainer. This belief is probably apocryphal, but elephants are capable of picking up objects as small as a coin. A 25-year-old female Asian elephant has been observed using her trunk with an amazing dexterity: she cracked peanuts with the back of her trunk, blew the shells away and ate the kernels.

The early nineteenth century French anatomist G. Cuvier and his colleagues examined the trunk of an elephant and estimated the number of muscles in it at about 40,000. A recent study has shown that the number of muscle units (fascicles) that manipulate this sensitive organ may total over 150,000, and that the trunk appears to have a more complex internal structure than once thought. It has no bones or cartilage; it is made of muscles, blood and lymph vessels, nerves, some fat, connective tissue, skin, hair, and bristles. The nostrils continue as separate openings from the base of the trunk to its tip. Each is lined with a membrane.

Measurements show that the trunk of an adult

Anthony Bannister/Oxford Scientific Films

Asian elephant can hold 8.5 liters (2.2 gallons) of water, and a thirsty adult bull elephant can drink 212 liters (56 gallons) of water in 4.6 minutes. Trunks are used for feeding, watering, dusting, smelling, touching, lifting, in sound production and communication, and as a weapon of defense and offense.

▲ *African elephants in mutual caress. This behavior strengthens social bonds between members of a herd.*

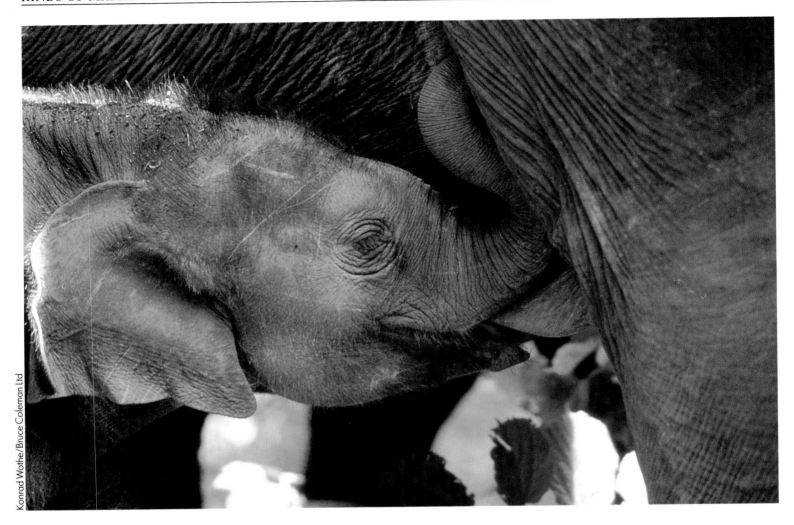

Konrad Wothe/Bruce Coleman Ltd

▲ *Asian elephant calf suckling: the trunk has to be lifted back so that its mouth can reach one of the mother's nipples. Gestation lasts about 20 months; the infant suckles for 8 to 10 months, after which the weaning process begins. Females reach sexual maturity when aged about 10 years but may not bear young until very much older.*

year. In her lifetime, which may be 50 to 80 years, a female elephant has the potential to give birth to 7 or 8 offspring under optimum environmental conditions This potential is rarely realized.

MUSTH

Musth is a periodical phenomenon known to occur in both Asian and African male elephants and is associated with physiological and behavioral changes. During musth, secretion oozes from the musth (temporal) gland, which is located beneath the skin midway between the eye and the ear on each side of the elephant's head. It is not found in any other living mammal. Musth does not occur in females although they do secrete fluid, but this fluid apparently differs in composition from that of adult males in musth.

Functions attributed to these secretions include sexual activity and communication. Some believe that musth is analogous to rut when males have a heightened mating drive and aggressiveness, and the females are receptive. However, musth does not necessarily involve heightened mating drive, nor are the females receptive when the males are in musth. In Hindi, *musth* means intoxicated. Elephants in musth can become uncontrollable, and captive elephants may kill their keepers.

ECOLOGY AND BEHAVIOR

Eighteen to twenty hours of an elephant's daily cycle is devoted to feeding or moving towards a food or water source: they consume 75 to 150 kilograms (165 to 330 pounds) of food and 83 to 140 liters (27 to 30 gallons) of water a day. Their diet is strictly herbivorous. Feeding constitutes about 80 percent of total behavior, and the balance is filled with activities such as bathing, playing, sleeping, and reproducing.

Elephants are highly social animals. The basic unit is a matriarchal family of five to ten animals; when families join they become a herd. The matriarch is the oldest and usually the most experienced female in the herd. During a drought, for instance, the matriarch will lead her family and relatives to the best possible foraging habitats. The rest of the herd learns and accumulates knowledge through close relationships; it is this tight bonding and herd experience, passed from one generation to the next, that has contributed to the survival of the elephant species.

Calves are tended not only by their mothers but also by other females in the herd. Males leave the herd at maturity, at about 13 years, and are sometimes joined by other males to form bachelor herds. Adult bulls join cow herds when a female is

AFRICAN & ASIAN ELEPHANTS: WHAT'S THE DIFFERENCE?

The most obvious difference between African and Asian elephants is the size of their ears — the African's are much larger. There are, however, many other differences; only some can be given here. African elephants are generally heavier and taller: a bull African can weigh 7 tonnes (14,000 pounds) and reach as much as 4 meters (over 12 feet). The maximum recorded weight and height for Asian elephants are 5.4 tonnes (12,000 pounds) and 3.35 meters (11¼ feet).

The African elephant has a concave back, whereas the Asian has a convex (humped) or level back. Tusks are usually present in both sexes of the African elephant; among Asians, mostly males possess them.

The trunk of the African elephant has more folds of skin in the form of "rings" or annulations, and the tip possesses two instead of one finger-like projection. The trunk of the African appears to be "floppy", while that of the Asian seems slightly more rigid. This can best be seen when they raise their trunks towards their foreheads.

Comments such as "Asian elephants are more easily trained than African" or "the trunk of the Asian is more versatile than that of the African" are partially or wholly correct, and can be explained in evolutionary context. The differences have

anatomical bases: the more advanced and specialized of the two species, the Asian elephant, has a higher degree of muscle coordination and therefore is able to perform more complicated antics than its cousin the African species.

Stephen J. Krasemann/NHPA

▲ African elephants. Note the large ears. African elephants are bigger than Asian elephants and both sexes develop large tusks.

Adrian Warren/Ardea London

Joanna Van Gruisen/Ardea London

▲ There are three subspecies of the Asian elephant, respectively from Sri Lanka, Indo-China, and Sumatra. This female from Sri Lanka lacks tusks and, like all Asian elephants, has an arched (sometimes straight) back; African elephants are sway-backed.

◀ A male Asian elephant from Malaysia. The tusks of this individual are of about average size for a mature male: the largest known tusks of this species were 3 meters (nearly 10 feet) long.

Gerald Thompson/Oxford Scientific Films

▲ *African elephants (above) feed largely on the leaves and branches of trees, particularly acacias. Asian elephants eat more grasses and shrubs. When food is scarce, African elephants push over trees to reach the topmost twigs.*

in estrus. There is no evidence of territoriality; home range is 10 to 100 square kilometers (4 to 400 square miles) or more, depending on sex, the size of the herd, and the season of the year.

Acuteness of the senses of elephants appear to change with age. Generally speaking, sight is poor, but in dim light it is good; hearing is excellent; sense of touch is very good; and taste seems to be selective. On balance, the African elephant feeds more on branches, twigs, and leaves than does the Asian elephant, which feeds predominantly on grassy materials. Elephants are crepuscular animals—that is, they are active mostly during early morning and twilight hours. When the sun is at its zenith and until early afternoon, elephants have their siesta.

The studies of K. B. Payne and her colleagues demonstrated the "secret" language of elephants. This infrasonic communication, a vocalization not audible to the human ear, appears to carry for long distances. Elephants that are located up to 2.4 kilometers (1½ miles) from the source of the sound seem to recognize signals emitted by other elephants and respond accordingly. Males move toward a sound produced by a female in estrus; other elephants at different locations seem to synchronize their behavior in response to stress signals.

JEHESKEL SHOSHANI

IVORY & CONSERVATION

In elephants, milk incisors or tusks are replaced by permanent second incisors within 6 to 12 months of birth. Permanent tusks grow continuously at the rate of about 17 centimeters (7 inches) a year and are composed mostly of dentine. Like all mammalian teeth, elephant incisors have pulp cavities containing blood vessels and nerves; tusks are thus sensitive to external pressure. On average, only about two-thirds of a tusk is visible externally, the rest being embedded in the socket within the cranium.

The longest recorded tusks of an African elephant measured 3.264 meters (10 feet 8½ inches); the heaviest weighed 102.7 kilograms (226½ pounds). Comparable figures for an Asian elephant are 3.02 meters (10 feet) and 39 kilograms (86 pounds). In 1990, a 45-year-old Asian elephant, Tommy, carried a pair of tusks measuring about 1.5 meters (5 feet) each, which are estimated to weigh at least 45 kilograms (100 pounds) each.

Elephants use their tusks to dig for water, salt, and roots; to debark trees; as levers for maneuvering felled trees and branches; for work (in domesticated animals); for display; for marking trees; as weapons of defense and offense; as trunk-rests; as protection for the trunk (like a car's bumper bar); and tusks may also play a role as a "status symbol". Just as humans are left or right handed, so too are elephants left or right tusked; the tusk that is used more than the other is called the master tusk. Master tusks can easily be distinguished since they are shorter and more rounded at the tip through wear.

In a cross section, a tusk exhibits a pattern of criss-cross lines that form small diamond-shaped areas visible to the naked eye. This pattern is unique to elephants. The term "ivory" should be applied only to elephants' tusks.

The hardness, and thus carvability, of ivory differs according to country of origin, habitat, and the animal's sex. Once removed from an elephant, ivory dries and begins to split along the concentric lines unless it is kept cool and moist. It also deteriorates if kept too moist; the water-absorbing properties of ivory are well known to certain African tribes, who use ivory as a rain predictor by planting it in the ground in selected locations.

As long as there is a demand for ivory, the slaughter of elephants in Africa will continue. Every year during the 1970s and 1980s, an estimated 70,000 elephants were killed to supply the insatiable demand for the white gold. Some of this ivory came from legal sources, such as culling operations, but about 80 percent of the tusks came from poached, brutally killed elephants. Poachers indiscriminately shoot males and females, young and old, and many

Silvestris/Australasian Nature Transparencies

◄ *African elephants in ritual combat. Such struggles seldom involve serious injury: an aggressive display by a dominant male is usually sufficient to discourage a rival with smaller tusks. The tusks of very old male African elephants occasionally grow so long and heavy that they become an encumbrance. At this age, however, the elephant has already passed its genes to the next generation.*

of the animals killed are in their reproductive prime. As poaching continues, the average size of tusks on the offspring of the surviving elephants is getting gradually smaller. In 1982 the average tusk weight of an African elephant was 9.7 kilograms (21 pounds); in 1988 the average was 5.9 kilograms (13 pounds). In Asia, where the size of tusks is much smaller in males and most females have no tusks, elephant poaching is almost non-existent.

Poaching is not, however, the only problem elephants face. The rapid growth of human populations leads to encroachment on wildlife habitat, so that elephants are 'compressed' into a given locale. In this limited area the vegetation is soon over-exploited, and damage to the habitat follows. Over all Africa in 1989, the wild elephant population was estimated at fewer than 750,000, which is half the number of ten years ago.

Various conservation measures are taken to protect wildlife in Africa, public education in the form of lectures, seminars, and travelling museums being the most important. Incentives have also been offered. For example, compensation is given to farmers whose crops are damaged or eaten by elephants. Protecting elephants may prove to be a very costly operation, yet it is imperative for the protection of other wildlife in the environment because, by preserving a large area for elephants to roam freely, one provides suitable habitat and protection for many other animal and plant species of the same ecosystem.

Because of the long generation time and continuous decline in numbers and habitat, elephants—Asian and African—cannot maintain healthy and reproducing populations. The present rate of elephant mortality is alarming. The Asian elephant *Elephas maximus* was declared an endangered species throughout its range in 1973, and is listed in Appendix I of the Convention on International Trade in Endangered Species of Wild Fauna and Flora (CITES). The African elephant *Loxodonta africana* was declared a threatened species and listed in Appendix II in 1978, and in 1990, it was upgraded to endangered species status. At the 1997 CITES meeting in Harare, Zimbabwe, it was adopted to transfer from Appendix I to II only the elephant populations of Botswana, Namibia, and Zimbabwe for limited and tightly monitored trade of ivory and hide.

To ensure the survival of wild elephants in Africa, four strategies are proposed: continue with public education; enforce existing laws; create natural corridors to connect isolated populations so that they can exchange genes; and establish cooperation among neighboring countries to ensure uninterrupted elephant habitats, especially where they are known to cross borders.

▼ *Ivory poaching became so intense in the latter half of the 1980s that there was a real danger of extinction of the species over portions of its range in East Africa. Since it has not been possible to control poaching, there is increasing international pressure to put an end to the ivory-carving industry, thereby reducing the market for tusks. This is, however, a controversial solution.*

Gerald Cubitt/Bruce Coleman Ltd

ODD-TOED UNGULATES

ORDER PERISSODACTYLA
• 3 families • 6 genera
• 18 species

SMALLEST & LARGEST

Mountain tapir *Tapirus pinchaque*
Head–body length: 180 cm (71 in)
Weight: 150–200 kg (330–440 lb)

White rhinoceros *Ceratotherium simum*
Head–body length: 370–400 cm (145–160 in)
Tail length: 70 cm (27½ in)
Weight: up to 2.3 tonnes (5,000 lb)

CONSERVATION WATCH
!!! The African wild ass *Equus africanus;* Sumatran rhinoceros *Dicerorhinus sumatrensis;* black rhinoceros *Diceros bicornis;* and Javan rhinoceros *Rhinoceros sondaicus* are listed as critically endangered.
!! The Grevy's zebra *Equus grevyi;* mountain zebra *Equus zebra;* mountain tapir *Tapirus pinchaque;* and great Indian rhinoceros *Rhinoceros unicornis* are listed as endangered.
! The vulnerable species are: Asian wild ass *Equus hemionus;* Central American tapir *Tapirus bairdii;* Malayan tapir *Tapirus indicus.*

The Perissodactyla represent the older of two orders of hoofed mammals or ungulates. They have retained functionally either three toes or a single toe on each limb, and hence are referred to as the odd-toed ungulates. The three families in the order are the Equidae (one genus), the Tapiridae (one genus) and the Rhinocerotidae (three genera). While the horses, asses, and zebras of the equids are clearly related, their association with the rare and strange tapir and the lumbering rhinoceros is more surprising.

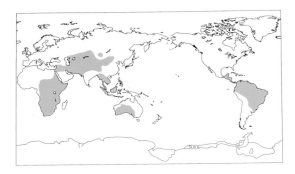

SPECIALIZED GRAZERS

Horses represent the group of ungulates that first adapted to running on the tips of their toes across the grassy plains that developed around 20 million years ago. Tapirs have remained forest animals, feeding on fruits as well as leaves. Rhinoceroses represented the dominant form of large herbivore between 25 and 40 million years ago, but they subsequently declined in diversity, although not necessarily in abundance. All these animals are distinguished from other ungulates not only by their number of toes, but also by the techniques they have evolved for digesting the cellulose content of plant cell walls. They are all specialized grazers and are found in grasslands in parts of Africa, Asia, and South America.

THE EQUID FAMILY

Horses and asses originated in North America, but the last of the American equids became extinct towards the end of the Pleistocene Ice Ages, about 12,000 years ago. The horse *Equus caballus* no longer exists in the ancestral wild form in Eurasia, having been domesticated by 2000 BC. Przewalski's horse *E. przewalski*, which is believed to be closely related to ancestral horses, persisted on the plains of Mongolia until recent times, but may now survive only in zoos. Domesticated horses have, however, established feral populations in all six continents. Two species of wild ass exist: the Asiatic ass *E. hemionus* in parts of the Middle East and southern Central Asia, and the African ass *E. africanus* in semi-desert areas of northeast Africa.

All equids are grazers, although they may include small amounts of leaves, twigs, and succulents in their diet during the lean season. All equids have a gestation period of about 12 months, so that they cannot produce a foal every year and still maintain synchrony with the seasonal cycle. As a result, reproductive rates are lower than in antelope and deer.

Horses

The original wild horses were more stockily built than the slender-limbed breeds prevalent today. Horses are the largest of the equids, weighing up to 700 kilograms (1,500 pounds). Unlike most horned ruminants, males are not much larger than females. This may be related to their fighting technique, which involves attempts to bite the legs of opponents and so depends more on agility than on strength.

The domestication of the horse was an important step in human cultural development, providing not only a beast of burden but also a vehicle for warfare. Foot-slogging armies had little answer to the mobility of attacking hordes mounted on their steeds, which led to the early conquests of European tribes by the Huns. Subsequently the near extermination of bison in North America and certain antelope in South Africa

▼ *Feral horses in New South Wales, Australia. Being descended from various breeds of domesticated stock that strayed or were let loose, these animals display great variability in color, size and form.*

Jean-Paul Ferrero/Auscape International

Liz & Tony Bomford/Ardea London

▲ *Przewalski's horse, from Mongolia, may no longer exist in a wild state. However, this stocky species, which is close to the ancestor of all the domesticated horses, is breeding well in an international program that involves many of the world's best zoos.*

was aided by the ability of hunters to ride in amongst herds on horseback, from where they could shoot large numbers of animals within a short time. Perhaps most African antelope were saved from the fate of American bison only because of the prevalence of horse sickness in much of tropical and subtropical Africa. Horses survive well without human care in the prairie region of North America as well as in the Camargue delta in France, while feral populations also exist in the Southern Tablelands of Australia, and northern and Central Australia, and in southwest Africa.

Asses

The donkey, or burro, is derived from the African wild ass. Used formerly as beasts of burden by gold prospectors, they have formed feral populations in Death Valley, California, and in the region of the Grand Canyon in Arizona. The African wild ass is now rare in the wild, persisting only in desert regions of Ethiopia and neighboring countries. They are shy and difficult to approach. Asiatic wild asses are somewhat more horse-like, and also quite rare. They are likewise desert inhabitants, dependent on their ability to outrun predators.

Jean-Paul Ferrero/Auscape International

▲ The African or Somali ass is closely related to the ancestor of the domestic donkey. Asses and donkeys are better adapted to dry conditions than are horses.

◄ A distinct species, the Asiatic ass exists in five subspecies, all of which are rare in the wild. The subspecies shown here is the onager.

Joanna Van Gruisen/Ardea London

▲ *Grant's zebra is a subspecies of Burchell's zebra. This plains-dwelling form is the only zebra that remains common.*

Zebras

Zebras are still abundant and widespread in Africa. There are three species: the plains or Burchell's zebra *E. burchelli,* which occurs throughout Africa in a number of races; Grevy's zebra *E. grevyi,* occupying northeast Africa; and the mountain zebra *E. zebra,* which has one race in Namibia and another in the southern Cape mountains. The latter two species have rather ass-like bodies.

Zebras are distinguished by their coloring of black stripes on a white background. Grevy's zebra and mountain zebra have finely divided patterns of black on white, while different races of plains zebra show varying degrees of brown shading between the stripes. A zebra with black stripes against a dull brown background, the quagga, formerly occurred in the Cape, but was exterminated by white settlers. Genetic evidence obtained from the hides of the two surviving museum specimens suggests that it may have been only an extreme color variant of the plains zebra. Weights vary from 235 kilograms (520 pounds) in some races of plains zebra to 400 kilograms (880 pounds) in Grevy's zebra. Zebras do not have much stamina, and early hunters found them easy to ride down on horseback. For the same reason, attempts to domesticate them were abandoned.

Grevy's zebra and mountain zebra occur in semi-desert areas, while plains zebra migrate through grassy plains and savanna regions.

Mountain zebra hooves have a rubbery texture to resist the bare rocks of arid ranges.

. Plains zebras commonly form mixed herds with various antelope. While they tend to be outnumbered in any local area by one or another species of antelope, these zebras are more widely distributed through habitat types and regions than any single antelope species. They occur in tight-knit groups of several mares accompanied by a stallion, as do feral horses and mountain zebras. Young females are abducted out of their family at the time they become sexually receptive. This is probably a mechanism to ensure outbreeding, in circumstances where stallions may remain attached to the same breeding group for many years. In contrast, in Grevy's zebra and wild asses, females form loose affiliations, while males occupy large territories.

Mixed herds may give zebras some additional degree of protection against predators. Adults commonly fall prey to lions, while hyenas and African wild dogs take a toll of foals. Stallions may actively try to protect their young against attacks by hyenas and wild dogs, while mares may even try to distract lions from killing foals. As non-ruminants, zebras are able to feed on the coarser, more stemmy fractions of grasses, including seed-bearing stems, so they do not compete strongly for food with grazing antelopes, even though they favor many of the same grass species.

Jen & Des Bartlett/Bruce Coleman Limited

▶ *The mountain zebra is now uncommon and regarded as vulnerable. The individual shown here, readily identified by the broad black and white stripes on the rump, was photographed in the extremely arid Namib Desert.*

TERRITORIAL SOCIAL SYSTEMS

Solitary animals may nevertheless have a highly developed social organization. For instance, white rhinos have a system of territories held by a proportion of the adult males. Each territory-holding male restricts his movements almost entirely to an area of about 2 square kilometers (¾ square mile). Within this area, he is supremely dominant over all other males. Neighboring territory holders express the balance of power in ritualized horn-to-horn confrontations at territory borders. Males scent-mark their territories by spraying their urine on borders and rhino trails crossing the territory, and also scatter their dung over the middens where other rhinos concentrate their feces. This means that any rhino entering a territory can soon detect that it is occupied, and perhaps even recognize the holder by his scent should there be a meeting. This does not necessarily deter other rhinos from entering. Females and young animals wander in and out of the territories. Adult males that have been unable to claim a territory of their own settle within the territory of another male and accept subordinate status by avoiding spraying their urine or scattering their dung. They noisily demonstrate their submission whenever they are confronted by the holder.

The significance of the territories becomes evident when a rhino female is sexually receptive. Territory-holding males attach themselves to such females and prevent them from crossing out of the territory by blocking their movements at boundaries. The result is that, when females permit courtship advances and eventually the lengthy mating, this takes place without interference from other males. Territories seem especially beneficial where a species attains high local densities and so many males are competing for limited mating opportunities. Territoriality is less clearly expressed

or is absent in other rhino species, which occur at lower densities than those typical of white rhinos.

Male Grevy's zebra and both species of wild ass also occupy territories, but these can be as large as 10 to 15 square kilometers (4 to 6 square miles). Perhaps in semi-desert environments animals can maintain visual surveillance over large areas, but stallions also mark boundaries with dung piles. Male territories are commonly found among African antelope species. Where female groups are highly mobile, as is the case in plains zebra and in horses, it is more effective for males to attach themselves to particular herds of females and accompany them in their movements. Hence the grouping patterns and mobility of the female segment of the population determine what type of social system males adopt.

▼ *All equids are social, and males engage in combat that is far from ritual. Grevy's zebra, shown here, is no more closely related to the other zebras than it is to horses or asses. We can regard zebras as horses that happen to be striped, and horses as unstriped zebras: stripes do not signify kinship.*

K.W. Fink/Ardea London

Kevin Shafer/Tom Stack & Associates

▲ Baird's tapir, from Central America, is a swamp-dweller. Like other tapirs, it forages for leaves and shoots with its flexible proboscis-like snout.

TAPIRS

Tapirs are squat animals, weighing about 300 kilograms (660 pounds), that are characterized by a short mobile trunk at the tip of the snout. There are three South American species: the Brazilian tapir *Tapirus terrestris*, the mountain tapir *T. pinchaque*, and Baird's tapir *T. bairdii*; and a single Asian species, the Malayan tapir *T. indicus*. While the American species are a dull brown, the Malayan tapir displays a contrasting black and white pattern. Newborn tapirs exhibit mottled cream flecks or stripes, evidently a camouflage for when they lie hidden while their mothers are absent.

The tapir's way of life has not changed much from that of ancestral forms of 20 million years ago. They browse forest shrubs and herbs, but also include much fruit and seeds in their diet when they are available. By feeding on large fruits, they may play an important role in disseminating the seeds of many forest trees. Tapirs seem to spend much time in water, and they are largely nocturnal and difficult to observe. They appear to be largely solitary in their habits. Male tapirs scent-mark their home areas by spraying urine. Because of the orientation of the flaccid penis, the urine is ejected backwards between the hindlegs (this feature is shared with rhinoceroses as well as with cats).

Tapirs are threatened by widespread logging and clearing of tropical forests in both Asia and South America.

All young tapirs are patterned with pale stripes and spots on a dark background. This juvenile Brazilian tapir blends marvellously with the background of dappled light in its woodland habitat.

▼ The Malayan tapir is very conspicuously marked with a disruptive pattern of black and white that provides effective camouflage in the dense rainforest where the few remaining animals survive.

182

RHINOCEROSES

Rhinoceroses are characterized by the horns they bear on their snouts. Unlike the horns of antelope, those of rhinos (as the animals are commonly called) lack a bony core. They consist of a mass of hollow filaments that adhere together and are attached fairly loosely to a roughened area of the skull. Early in their evolutionary history rhinos developed a tendency towards large size. An extinct hornless form, *Indricotherium,* attaining 5.5 meters (18 feet) at the shoulder and weighing perhaps 20 tonnes (over 40,000 pounds), was the largest land mammal ever. Like elephants, male rhinos have no scrotum, the testes remaining internal. Five species of rhinoceros have survived. They fall into three distinct subfamilies, one restricted to Africa and the other two to Asia, although they formerly extended into Europe.

African rhinoceroses

The African two-horned rhinos are represented by two species, the black or hook-lipped rhinoceros *Diceros bicornis* and the white or square-lipped rhinoceros *Ceratotherium simum.* Although assigned to separate genera, the two species have a common ancestor that existed about 10 million years ago. They are distinguished most importantly by their feeding habits: the white rhino is a grazer, with a long head and wide lips designed for cropping short grasses, and the black rhino is a browser, with a prehensile upper lip for drawing branch tips into its mouth. There is in fact no clear distinction in skin color between the two species, despite their popular names. Both are basically grey, but with the precise tinge modified by local soil color. The name white rhino has been ascribed to a corruption of a Cape Dutch name, *wijd mond* (wide mouth) rhinoceros, but this is a myth. Probably the first white rhinos encountered by European explorers in the northern Cape had been wallowing in pale, calcium-rich soil.

The white rhino is by far the larger of the two species, with adult males reaching weights of about 2.3 tonnes (5,000 pounds). It is regarded as the third-largest land mammal alive today, after the two species of elephant. Other distinguishing features include the shape of the neck, back, ears, and folds on the skin. The white rhino occurs in two distinct races. The southern form *C. s. simum* was distributed through southern Africa south of the Zambezi River. The northern race *C. s. cottoni* was found west of the upper Nile River in parts of Sudan, Zaire, and neighboring countries. Differences in appearance between the two races

◄ The white rhinoceros is only a little paler than its "black" (actually gray) African relative. It is better referred to as the square-lipped rhinoceros, in reference to its wide, straight, non-hooked upper lip. It is the only grazing rhinoceros.

▼ The black rhinoceros is better referred to as the hook-lipped African rhinoceros, in reference to the short prehensile projection of the upper lip, which is used to pull leaves into the mouth. Like the rhinoceroses of Asia, it is a browser.

Rod Williams/Bruce Coleman Limited

▲ *The Sumatran rhinoceros has a sparse covering of long hair. In this respect, and in the retention of incisor and canine teeth, it is more primitive than the African rhinoceroses.*

The white rhino is mild and inoffensive in temperament and was soon hunted toward extinction in southern Africa when guns capable of piercing its thick hide became available. The species is more social than other rhinos, forming groups of three to ten made up of young animals and females without calves. Although these groups, and females accompanied by calves, move independently, they approach one another to engage in playful horn wrestling when they meet. Adult males are solitary and territorial.

The black rhino attains weights of up to 1.3 tonnes (nearly 3,000 pounds). It is renowned for its aggressive charges at human intruders and their vehicles, making puffing noises that sound like a steam engine. This behavior, coupled with the thick bush the animals commonly inhabit, proved fairly effective deterrents to human hunters until recently. Hence the species was widely distributed through much of savanna Africa, from the Cape to Somalia, and even into desert regions in Namibia and northern Kenya. Various races have been distinguished. The large Cape form *D. b. bicornis* has become extinct in the Cape, but may still be represented by Namibian animals. A smaller southern form *D. b. minor* has a distribution

are small, but they occur in different kinds of savanna and appear to be fairly distinct genetically. The existence of the northern race was only confirmed this century, such is the remoteness of the region. Fossil remains indicate that the white rhino was once more widely distributed through East and North Africa.

Francisco Erize/Bruce Coleman Limited

Alain Compost/Bruce Coleman Limited

extending from the eastern parts of South Africa through Zimbabwe and Zambia into southern Tanzania. In Kenya it is replaced by another form, *D. b. michaeli*. Other races may occur in Somalia and in West Africa, but too few specimens remain to confirm their distinctiveness. In some regions black rhinos are characterized by peculiar festering sores behind their shoulders. These are caused by a filarial roundworm, which lives in the wounds. White rhinos never develop these sores. The two species of African rhino usually ignore one another when they meet, but sometimes appear mildly curious about each other, and may even approach to rub horns.

Asian rhinoceroses

The Asian two-horned rhinos are represented by a single species, the Sumatran rhinoceros *Dicerorhinus sumatrensis*. This is a small (for a rhino) and somewhat hairy animal, which weighs up to 800 kilograms (1,750 pounds). It is a browser occupying mountainous forests not only in Sumatra, but also in other parts of Southeast Asia. Other rhinos of this subfamily occurred in Europe and Asia during the period of the Ice Ages. They included the famous woolly rhinoceros *Coelodonta antiquitatis*. Specimens preserved in ice

in Siberia confirm that it was a grazer, like the white rhino.

The Asian one-horned rhinoceroses are represented by the great Indian rhinoceros *Rhinoceros unicornis,* which rivals the white rhino in size, and by the smaller Javan rhinoceros *R. sondaicus.* Both bear a single horn on the tip of the snout. A prominent characteristic of the Indian rhino is the armor-like folds on the skin, which feature in a famous tale by Rudyard Kipling. The Indian rhino does not use its horns in fighting, as the African rhinos do, but lunges at opponents with its tusk-like incisor teeth. The Indian rhino is largely a grazer, although not as exclusively as the white rhino. It favours tall swampy grasslands in northern India and adjacent Nepal, and spends much time wallowing in ponds. Although animals may congregate around ponds, they are largely solitary in their movements. Indian rhinos exhibit noisy mating chases, with females making honking sounds and males squeaky panting noises. They are famous for their prolonged matings, copulation commonly persisting for as long as 60 minutes. The Javan rhino is a browser occupying lowland forests, formerly occurring through much of Southeast Asia. Little is known of its behavior, and there are no specimens in zoos.

▲ *Much smaller than its Indian relative but also with a very thick skin and a single horn, the Javan rhinoceros has a projecting, prehensile upper lip, which is used to grasp leaves and twigs. It is often encountered in water.*

◀ *The Indian rhinoceros has such extremely thick skin that it is arranged in plates, like a jointed suit of armor.*

THE TRADE IN RHINO HORNS

Rhinos of all species are being pushed towards extinction on account of the very high prices fetched by their horns. These horns are used primarily as a medicine in China and neighboring countries of the Far East. Horns are ground into a powder and swallowed as a potion. They were also taken as an aphrodisiac in a region of west India, but this practice is no longer of much importance. Another use is for making dagger handles, which are especially prized as a symbol of manliness in North Yemen. In fact the substance of rhino horn, a protein called keratin, is no different from that making up the outer covering of the horns of cattle and antelope, hooves, and human fingernails.

Although the efficacy of rhino horn as a drug may be quite mythical, the fact remains that horns fetch extremely high prices (exceeding that of gold per unit weight), and the value tends to rise as rhinos become rarer. This spurs increasing endeavors by unscrupulous entrepreneurs to acquire horns, paying peasant farmers prices to hunt the animals that are vastly in excess of the honest wages these people could otherwise earn. The result is that conservation agencies in Africa and Asia have been fighting a losing battle to protect the vanishing remnants of the herds of these great beasts.

There is no easy solution to this problem. As horns will regrow if removed, some conservationists advocate farming rhinos by repeated cropping of their horns, thereby producing a steady supply of revenue to impoverished communities through honest channels. However, the relationship between remaining numbers of rhinos and potential markets in the east is such that the temptation to hunt will continue, forcing rhino populations ever downwards. Another proposal, already instituted in Kenya, is for rhinos to be handed over to private owners who have the wealth for the fencing and patrolling to protect the animals. A third suggestion is for as many of the remaining rhinos as possible to be moved into zoos, there to be bred awaiting more favorable circumstances for them to be reintroduced into the wild. However, by the time

Anthony Bannister/NHPA

memory of the illegal horn trades has faded, the rhinos' former habitats may have become settled by peasant farmers desperately trying to eke out an existence. Some conservationists believe that every attempt should be made to keep rhinos in the wild, however hopeless the situation may seem.

Rhinos can serve as potent symbols for marshalling international aid for parks and other conservation action in impoverished Third World countries. While the value of rhinos to humans today may be largely symbolic, they may serve to gain protection for many lesser species that are also threatened but attract little notice. The ultimate value to humankind of these inconspicuous species, such as plants and insects, may be vastly greater than that of spectacular mammals like rhinos.

▲ *Horns of the black rhinoceros, confiscated by African wildlife authorities. It has been seriously suggested that the last of the world's black rhinoceroses might be saved by government programs of de-horning the animals, making them valueless to poachers.*

▶ *Symbols of death. Daggers with rhinoceros-horn shafts on sale in North Yemen.*

CONSERVATION

Rhinos have been the center of much attention from conservationists. The Javan rhino is among the rarest species alive today, with a surviving population of about 60 animals confined to a single reserve at the western tip of Java. The Sumatran rhino is only slightly better off, as the 800 or so animals remaining are thinly scattered through Malaysia and adjacent countries, and are vulnerable to both poaching and logging. Remnants of the Indian rhino are protected in reserves in northeast India and Nepal, but number under 2,000. The southern white rhino was reduced to perhaps 50 animals in the Umfolozi Reserve in South Africa at the turn of the century, but under careful protection its number has increased steadily to over 4,000 now. This has enabled the species to be reintroduced into many areas from which it had been exterminated by early white hunters, for example in Zimbabwe, Botswana, and Mozambique. The northern white rhino maintained its abundance until the late 1960s, but has now been almost exterminated. Only about 20 animals remain, all confined to the Garamba National Park in northern Zaire. Regular monitoring of these few individuals has halted

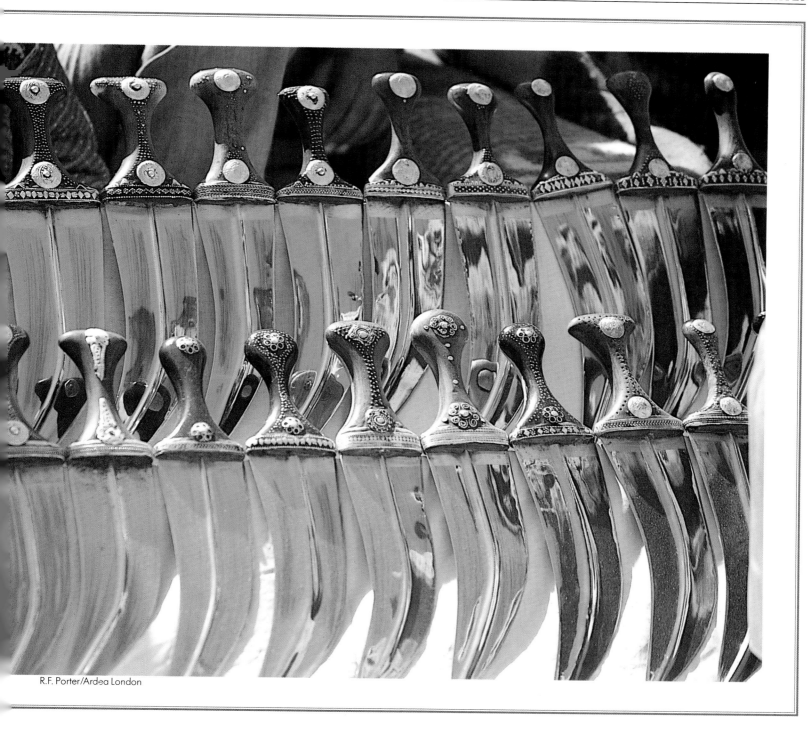

R.F. Porter/Ardea London

further losses. The black rhino was formerly the most abundant and widespread of the five species, with a total population of over 60,000 in 1970. This situation has changed drastically in recent years: widespread poaching has caused its numbers to plummet to perhaps 3,500 at present.

The decline in the abundance of rhinos is entirely due to humans, through hunting coupled with habitat destruction. Adult animals are large and formidable, and hence rarely fall prey to carnivores, although young animals are vulnerable. Hence adult mortality is low and is balanced by slow reproductive rates. The gestation period is 16 months in white rhinos and Indian rhinos, and 15 months in black rhinos. The birth interval is about two and a half to three years, so that the maximum rate of population growth is under 10 percent a year. Once levels of the harvest by human hunters exceed this threshold, populations are driven downwards. This happened to the white rhino between 1860 and 1890, when an abundant and widespread species was reduced to a few score animals. Today unscrupulous hunters are being lured by the high prices fetched by rhino horns in markets in Asia and the Middle East.

NORMAN OWEN-SMITH

HYRAXES

SIZE

Hyraxes (family Procaviidae)
Head–body length: 30–60 cm
(12–24 in)
Tail length: (not visible) up to
3 cm (1⅕ in)
Shoulder height: 30 cm (12 in)
Weight: 1.2–5.5 kg (2⅖–12 lb)

CONSERVATION WATCH
! 2 species—eastern tree hyrax
Dendrohyrax validus and
Heterohyrax brucei—are listed
as vulnerable.

S ometimes mistaken for rodents or rabbits, hyraxes or dassies are small but robust
mammals that resemble a huge guinea-pig. However, hyraxes are really so
different from all other mammals that zoologists place them in an order by
themselves — the Hyracoidea. This order has one extant family, Procaviidae, which
includes three living genera: *Dendrohyrax,* the tree hyrax, *Heterohyrax,* the yellow-
spotted rock hyrax, and *Procavia,* the rock hyrax. The classification of *Procavia* is still
provisional: there are at least four species — more, according to some zoologists.

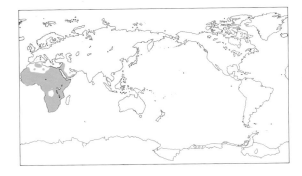

ROBUST ROCK-DWELLERS
The earliest fossil hyracoids date to the Eocene
epoch, about 50 million years ago. Living hyraxes
are all approximately rabbit-sized, but extinct
species reached the size of a tapir. It appears that
hyraxes, extinct and extant, were adapted to a
variety of ecological herbivorous niches, though
today they are restricted to tropical and
subtropical habitats, in altitudes from 400 meters
(1,300 feet) below sea level, at the shores of the
Dead Sea, to mountains 3,800 meters (12,500
feet) above sea level in East Africa.

For their size, hyraxes are robust, stocky
animals. They have a muscular, massive short

neck; a long and arched body, stubby legs, and a
tail is not visible. They have large eyes, medium-
sized ears, and a truncated snout with a cleft in the
upper lip. Sexual dimorphism is evident: the males
are heavier, more muscular, possess longer upper
incisors, and behave more aggressively than the
females. The coat is dense and consists of short
underfur and long tactile guard hairs. General body
color is highly variable among the genera and
species, from light gray or yellowish brown to dark
brown; the flanks are lighter and the underparts
are buff. In most species, there is a light, thin band
of hair along the dorsal edge, and a tuft of long hair
a different color from the surrounding area is
present at about the center of the back. This tuft
covers a scent gland that secretes a sticky, smelly
substance, believed to have a communicatory
function. The skin is relatively thick.

Hyraxes walk on the soles of their feet, which are
unique, with large, rubbery-soft elastic pads which
are kept moist by secretory glands; these
adaptations assure them a firm grip on rocks and
trees. The forefoot has five digits but only four are
functional, while the hindfoot has four digits with
three functional. Except for the inner toe of the
hindfoot, all digits terminate in short, flat, hoof-
like nails. That of the inner hind toe is a long,
curved, claw-like nail, which is used for grooming.
Grooming is also accomplished by the spatula-
shaped lower incisors in combination with the
upper teeth. A hyrax grooms itself, and can usually
reach most parts of its body, but occasionally
hyraxes engage in social grooming.

REPRODUCTION AND TERRITORY
An animal the size of a hyrax (for example, a
rabbit) usually has a gestation period of about one
month, but pregnancy for hyraxes has been
reported to last from six to eight months. One
possible explanation for such an extremely long
gestation period is that the hyrax is an evolutionary
vestige of much larger ancestral animals.

During the mating season, *Procavia,* which mark
rocks with their dorsal scent glands, may exhibit
territoriality-related behavior. The odor may also
serve as a form of social communication among the
colony members and help the young identify or
locate their mothers. The mating season in the wild
varies among the genera, but most have their

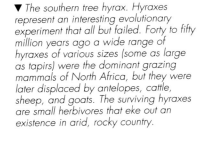

▼ *The southern tree hyrax. Hyraxes
represent an interesting evolutionary
experiment that all but failed. Forty to fifty
million years ago a wide range of
hyraxes of various sizes (some as large
as tapirs) were the dominant grazing
mammals of North Africa, but they were
later displaced by antelopes, cattle,
sheep, and goats. The surviving hyraxes
are small herbivores that eke out an
existence in arid, rocky country.*

John Shaw/NHPA

young during the spring season. *Dendrohyrax* usually produce one or two offspring and *Procavia* and *Heterohyrax*, two to four. The newborn are highly precocial, like all living ungulates. Young *Procavia* often climb on their mothers' backs, engage in "play" behavior, and are more active than adults. Sexual maturity is attained at one year.

GREGARIOUS VEGETARIANS

The hyrax life span is about 10 years and a record of 14 years has been noted. Colonies of up to 80 rock hyraxes consist of a few to several families, each of which is headed by a male. All species are vegetarians, but they sometimes eat invertebrates. Rock hyraxes feed on the ground as well. Usually, one or more individuals keep watch and warn other members of their group of an approaching predator or potential predator with a loud warning cry. All the animals suddenly disappear among the rocks and boulders, but once the danger is over, they gradually reappear and resume their activities. Natural predators of hyraxes include leopards, wild dogs, eagles, and pythons, while the indigenous people of some African countries hunt them for their flesh and skins. In places where their predators have been reduced or eliminated, hyrax numbers increase and they become a nuisance.

Most hyraxes are active during the morning and late afternoon, when they bask in the sun—often close together in a group—and feed. The basking, passive behavior presumably increases their metabolic rate and prepares them to forage. Hot hours of the day are spent in the shade. Hyraxes have a habit of staying in one spot and staring in one direction for a long time.

Hyraxes do not burrow; crevices and spaces

Peter Davey/Bruce Coleman Ltd

among rocks provide them with most of the shelter they need. On occasion they have been known to inhabit the burrows of aardvarks and meerkats.

No other animal sounds like a hyrax. Their vocabulary changes and increases throughout life; the young ones utter long chatters which increase in their intensity from beginning to end of each bout. It has been reported that *Procavia* infants make only five of the twenty-one sounds made by adults. The best known calls are those of *Dendrohyrax*. They begin to vocalize soon after dark in a series of croaks that end in a loud scream. Travelers on safari in Africa are often warned of these sounds and told not to confuse them with those of bandits.

JEHESKEL SHOSHANI

▲ Yellow-spotted rock hyraxes on an old termite mound. Although they look like guinea pigs, among living mammals hyraxes are most closely related to elephants, sea cows, and the aardvark. They live in groups with a complex social structure, ever on the alert for predators. They are active during the day, which is unusual for small mammals.

S. Krasemann/NHPA

◄ Yellow-spotted rock hyraxes basking. Living in a habitat with sparse food, hyraxes are adapted to a low-energy diet. They are unable to regulate their body temperature effectively but they conserve heat by huddling together and warm themselves by basking in the sun.

AARDVARK

ORDER TUBULIDENTATA
• 1 family • 1 genus • 1 species

SIZE

Aardvark *Orycteropus afer*
Head–body length: 100–160 cm
(40–65 in)
Tail length: 44–80 cm (17–30 in)
Shoulder height: 58–66 cm
(23–26 in)
Weight: 40–100 kg (88–220 lb),
usually 50–70 kg (110–154 lb)

CONSERVATION WATCH
■ Not endangered, but sanctions
apply to trade in their products.

The aardvark—a name derived from Dutch, through Afrikaans, for "earth pig"—is one of the strangest of all living mammals. It is classified in an order of its own, the Tubulidentata. The order consists of one family, Orycteropodidae, with four extinct and one extant genera. The only surviving species is the aardvark *Orycteropus afer.* Tubulidentata probably originated in Africa at some time in the Paleocene, about 65 million years ago, or earlier, and are believed to have evolved from an early hoofed-type mammal. Because they exhibit primitive mammalian anatomical and molecular characters, aardvarks can be thought of as "living fossils".

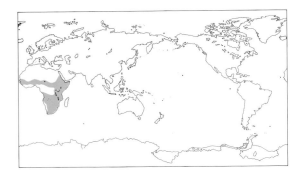

FIERCE AND PIG-LIKE

Although aardvarks eat ants they are not "ant-eaters" in the strictest zoological sense. Early naturalists classified aardvarks together with pangolins and armadillos, but as it became evident that the similarities between them are due to convergence rather than affinities, this view was abandoned and today each one of these groups is assigned to a different order.

The aardvark is a medium-sized pig-like animal with a stocky body, a short neck, and an arched back; it has a long snout, large ears resembling those of a donkey, and a long muscular tail that is

thick at its base. The skin is thick and sparsely covered with bristly hair, light in color on the body (though usually stained by soil) and darker on the limbs. The head and tail are whitish. Females are slightly smaller than males; their heads are a lighter color and the tail has a bright white tip.

The snout contains a labyrinth of nine to ten thin bones — the highest number of all mammals. The nostrils can be closed by means of muscular contraction of hair, an adaptation that prevents ants and termites from entering the snout. The eyes are reduced and the animal is color blind, but hearing and smell are acute. The mouth is small with a long, narrow tongue; food particles are taken in with the tongue and scraped off by ridges on the palate as the tongue protrudes again. The jaw muscles are weak and the mandible is slender.

An aardvark's forelimbs are short, powerful digging tools that have four shovel-shaped claws on each foot. The hindlegs, which have five clawed toes, are long and function as support and as a springboard for the body. With its massive skeleton, thick skin, and sharp claws, an aardvark is well armed — humans and hyenas are among the few predators that will attempt to hunt a healthy adult. The young, the old, and the sick, however, are preyed on by lions, leopards, cheetahs, hunting dogs, and pythons. When attacked, aardvarks are fierce and will kick and slash with their legs and claws. With great speed they can somersault and stand on their hindlegs and tails to defend themselves with their forefeet.

FEEDING AND NESTING

Unlike those of other mammals, the teeth of aardvarks are oval or figure-of-eight-shaped, flat on the top, and columnar. They consist of hexagonal prisms with many fused minute dentine columns and pulp cavities that appear as tubes on the chewing surface of the tooth. The teeth grow continuously, are without enamel, are covered with cement, and are rootless. Adult aardvarks have no incisors or canines, and only eight premolars and twelve molars. The cheek teeth can easily crush the hard exoskeletons of ants and termites, which constitute their main diet. Aardvarks also eat locusts, and grasshoppers and insects of the family Scarabidae. The stomach is one-chambered, and

▼ The aardvark feeds on ants and termites by digging into their nests wtih its powerful, strongly clawed forelegs. Insects are picked up on a long sticky tongue and swallowed with little or no chewing. The nostrils, on the tip of the snout, can be closed to prevent insects entering.

Gary Milburn/Tom Stack & Associates

the cecum is large for an insectivorous mammal. Aardvarks have been observed eating vegetable matter, especially the fruits of a wild plant known in South Africa as "aardvark cucumber", apparently for its moisture.

Generally speaking, wherever termites are found one can expect to find aardvarks: open canopy forests, bush veldts, and savannas are among their favorite habitats. An aardvark can excavate a termite mound in a few minutes where a human would need to use an ax or other heavy-duty tool. Termites and ants are ingested by means of a sticky, 30-centimeter (14-inch) extensible tongue. Most termites and ants are unable to hurt aardvarks, who are protected by a thick hide. On one occasion, however, I observed a female aardvark that had been stung by large ants; she rolled about in a frenzy for three minutes, rubbing her body against the hard ground, rocks, and logs.

While foraging, the animal moves in a zigzag path and continuously sniffs the ground in an area about 30 meters (33 yards) wide. It may travel 10 kilometers (6 miles) or even up to 30 kilometers (18 miles) a night in search of food.

Aardvarks dig three types of burrows: for food, for temporary shelter, and for permanent residence. Food and temporary burrows vary from shallow to deep holes excavated by the animals, but they sometimes use abandoned termite mounds. A permanent residence is about 2 to 3 meters (2 to 3 yards) long, usually with one entrance descending at an angle of about 45°. It may be a single main straight tunnel or an extensive tunnel system with many access holes. Sleeping chambers are wide enough for the animal to turn around: aardvarks enter and leave their burrows head first. Unused aardvark burrows are important for the survival of many small species such as hyraxes, which occupy them as dens or refuges in case of fire.

REPRODUCTION AND TERRITORY

The uterus of a female aardvark is similar to that found in rodents and rabbits. Nipples are in two pairs, one pair in the lower abdomen and one in the groin; the number of milk ducts varies from one to three. The breeding season appears to relate to latitude: in habitats away from the Equator breeding begins earlier than in habitats close to the Equator. Births occur from May to July in South Africa and Ethiopia, and later in the year in Zaire and other equatorial countries. Estrus is signalled by swelling of the vagina, and sometimes a discharge is visible.

The offspring, which usually weighs about 2 kilograms (4 pounds 6 ounces) and measures 55 centimeters (22 inches), is born after a seven-month gestation period. The newborn are partly precocial: their eyes are open and their claws are well developed. They join their mothers on nocturnal foraging trips at the age of two weeks and remain

Clem Haagner/Ardea London

with them until the next mating season. Sexual maturity is attained at about 2 years. In captivity, a female has given birth to 11 young in 16 years, and a male has sired 18 offspring by the age of 24.

Males are more vagabond than females and spend most of the year in separate burrows, except during the breeding season. Evidence for territoriality is inconclusive: in areas where density is high, several animals may occupy and feed in the same or overlapping home ranges. Aardvarks visit watering holes frequently and are good swimmers.

AARDVARKS AND HUMANS

Aardvarks perform an important function by controlling termites. In areas where aardvarks and other insectivorous animals have been exterminated, crops have suffered extensive damage. Grazing by wild and domestic ungulates creates good conditions for termites, which in turn are eaten by aardvarks; but aardvark burrows damage farmland and are hazards to vehicles and galloping horses. Bushmen, Hottentots, and other residents still hunt aardvarks for their meat and hide, as well as for medicines and amulets and for sport. The meat is said to be very tasty, similar to that of a pig.

Aardvarks are rarely observed in the wild. They are nocturnal, solitary, and elusive and their habitat is shrinking as more land is cultivated.

JEHESKEL SHOSHANI

▲ During the day, an aardvark usually sleeps in its burrow: it is rare for one to be active by day, as in the photograph above. Although it has not been proved, it seems likely that the large ears act as radiators that assist in temperature regulation, as in other desert dwellers such as the fennec, jack rabbit, and bilby.

Lesser mouse deer *Tragulus javanicus*
Head–body length: 44–48 cm (17–19 in)
Weight: 1.7–2.6 kg (3¾–5¼ lb)

TALLEST Giraffe *Giraffa camelopardalis*
Head–body length: 4.2 m (13¾ ft)
Height to horn tips: 5 m (16½ ft)
Weight: 1.35 tonnes (3,000 lb)

MOST MASSIVE Hippopotamus
Hippopotamus amphibius
Head–body length: 3.4 m (11¼ ft)
Weight: 2.4 tonnes (5,300 lb)

CONSERVATION WATCH
!!! 9 species are listed as critically endangered: Visayan warty pig *Sus cebifrons*; pygmy hog *Sus salvanius*; Pere David's deer *Elaphurus davidianus*; kouprey *Bos sauveli*; Walia ibex *Capra walia*; Hunter's antelope *Damaliscus hunteri*; Queen of Sheba's gazelle *Gazella bilkis*; scimitar-horned oryx *Oryx dammah*; Przewalski's gazelle *Procapra przewalskii*.
!! 28 species are endangered, including: Chacoan peccary *Catagonus wagneri*; wild Bactrian camel *Camelus bactrianus*; Philippine spotted deer *Cervus alfredi*; European bison *Bison bonasus*; wild water buffalo *Bubalus bubalis*; markhor *Capra falconeri*; slender-horned gazelle *Gazella leptoceros*; dwarf blue sheep *Pseudois schaeferi*; mountain nyala *Tragelaphus buxtoni*.
! 35 species are listed as vulnerable.

EVEN-TOED UNGULATES

The Artiodactyla, or cloven-hoofed mammals, are the younger of two orders of hoofed mammals, or ungulates. They have two or four weight-bearing toes on each foot, hence the name even-toed ungulates. Artiodactyls are a successful group of herbivores: they have a high species diversity; they inhabit a wide range of habitats, from tropical to polar; and they have a large biomass and wide geographic distribution, occurring naturally in all continents except Australia and Antarctica.

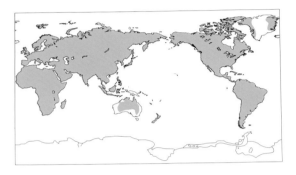

RUNNING TO SAFETY

The Artiodactyla include non-ruminating forms such as the pigs of the Old World (family Suidae); the pig-like peccaries of the New World (Tayasuidae); and hippos, now restricted to Africa (Hippopotamidae). The ruminants include camels (Camelidae); the chevrotains or mouse deer (Tragulidae); the musk deer (Moschidae); antlered deer (Cervidae); giraffes (Giraffidae), two species, now restricted to Africa; the lone American pronghorn (Antilocapridae); and numerous hollow-horned bovids (Bovidae).

A series of new species, previously unknown to science, have been found in Southeast Asia and South America. In South America, new genetic species of brocket deer are being revealed by DNA analyses. In Laos and Cambodia, the soala *Pseudoryx nghetinhensis*, the giant muntjac *Megamuntiacus vuquangensis*, and the Annamite muntjac *Muntiacus annamitensis* have been discovered. Another form, of which only the horns and hide are known, the Linh-guong *Pseudonovibos spiralis* remains to be taken alive. Incidentally, the Chacoan peccary was discovered alive only in 1974, although its bones had been known from the fossil record.

Artiodactyls are herbivores, though some pigs are omnivores, and all are prey of carnivores. Consequently, their biology is dominated by means of evading predators. The foot structure of the artiodactyls, evolved for sprinting, reflects this. The bones of the soles of their feet are large, to absorb the stress of running, and fuse into a

common cannon bone in all but the pigs, hippos, and the front legs of the peccaries. The hooves, or large claws in the case of camels, ensure immediate traction on almost any earth surface when power is suddenly and massively applied. Only the hippos use water to defeat predation, and so they differ in build from other artiodactyls.

While some pigs have simple stomachs, the majority of artiodactyls have complex stomachs and enlarged ceca that serve in the anaerobic fermentation of plant food. Even some pigs, and especially the peccaries and hippos, have complex stomachs. Fermentation allows a percentage of the otherwise indigestible cellulose in the cell walls of plants to be digested into metabolizable products, namely short-chain fatty acids. These are directly absorbed into the bloodstream through the walls of the stomach. The fermentation of plant food and the absorption of fatty acids is followed by the digestion of the bacteria and associated organisms in the true stomach. That is, ruminants culture their own protein source. This not only supplies the protein, but also most of the vitamins as well.

The fermentation vat also accepts inorganic matter, such as sulfur, which the bacteria incorporate into their bodies, creating sulfur-bearing amino acids, so vital for growing connective tissue and hair, horns, and hooves. For this reason mineral licks are sought out not only for their vital minerals, such as magnesium, sodium, and calcium, but also for sulfur. In fact, in the beautiful white Dall's sheep, of northern Canada and Alaska, females and young make a mineral lick the center of their summer activity, for as long as the females are heavily lactating.

EVOLUTION

The artiodactyls appeared early in the age of mammals, in the Eocene (57 to 37 million years ago), as small, forest-adapted omnivores. They were barely the size of rabbits and, like them, apparently depended for survival on hiding plus a swift getaway by hopping rapidly along the ground and over obstacles. Proceeding by leaps and bounds is costly in energy, as the runner must

generate considerable lift with each jump, and therefore tires rapidly. However, it must be an effective way to escape predators, because for small ungulates it is still the primary mode of escape.

In tropical forests only a small amount of easily digestible food, such as fallen fruit or shoots growing on the ground, is available to ground dwellers, so the density of such herbivores is low. Artiodactyls blossomed into great diversity during the middle of the Tertiary (65 to 2 million years ago), when savanna and steppe replaced much of the earlier forests. The artiodactyls now had access to grasslands which, unlike tall forests, could be totally exploited for food. With that arose a multitude of highly gregarious, large-bodied

runners that lived in high density on open plains.

Gregariousness and speedy, enduring running leave their mark on body shape and weapons. The chest cavity enlarges to house large lungs and heart; the shoulders become more muscular; and the legs equalize in length and become slim and light. Some species—such as saiga antelopes, reindeer, and addax—have become "short-legged" runners that make their escape across level, even, hard, unobstructed ground. They run with their heads low and minimize costly body lift, converting almost all their energy into forward propulsion.

Then there are the "long-legged" runners, which run with their heads held high. They are specialized to run over uneven, broken terrain,

▲ *Giraffes run with a "pacing" gait, both legs on each side moving more or less together. The forelegs are very long and must be spread apart to allow the mouth to be brought low enough to drink.*

Hans Reinhard/Bruce Coleman Ltd

▲ *The babirusa (top, center), unlike the other pigs, has tusks that seem to be purely ornamental. In the males, they usually jut through the top of the snout. The red river hog (above, right) is probably the most colorful of the pig family. It has several color forms, not all of them red. The pig-like chacoan peccary (above, left) was discovered in 1975. It is the largest and rarest of the three species of peccary.*

where they need an early warning of obstructions to come and the ability to shift their hooves quickly within each bound to avoid obstacles. Examples are the dama gazelle, the argali sheep, and the pronghorn. And there are runners that specialize in throwing a stream of obstacles into a pursuing predator's path. Examples are the mule deer, with its unique bounding gait, or the moose that trots smoothly and fast over low obstacles that pursuing bears or wolves must cross in costly jumps.

ATTACK AND DEFENSE

Not only the body form changes with gregariousness; so do weapons. In bounding runners from tropical thickets, the weapons are long combat teeth, and short and sharp horns, or outgrowths on the head that allow the head to be used as a club. These types of weapons are associated with the defense of material resources within a territory. Weapons that damage the body surface, and which will not stick in an opponent's body, combined with tactics of surprise attacks, are highly adaptive. Such attacks aim at quickly provoking great pain, making the intruder run away with a most unpleasant memory of the event and the locality. Sharp pain discourages retaliation that would very likely damage the attacking territory-owner. So-called "knives and daggers" are universally associated with the defense of resource, but not of mating, territories.

In the so-called "selfish herd" on open plains, damaging weapons are a liability. By grouping in the open, each individual gains from the presence of the others: the more there are, the more secure each individual. A predator striking may kill one, and the chances of being that one decline with companions. When fleeing, an animal has a much better chance in a large than in a small group of not being last: it's the last one that gets caught! By running first and attracting others to follow one is pretty sure of not being the last. Consequently, it pays to have a "follow me" marker on the tail. That appears to be the origin of the large, species-specific rump patches in gregarious species.

Weapons must allow the animal to defeat an opponent, but their use must not normally lead to retaliation or wounding of either participant in a fight. Blood attracts predators and it matters little whose blood it is. Consequently, with gregariousness a new type of weapon arises: the

horns or antlers change into twisted, rugose, or branched "baskets" that readily catch and bind the opponent's head, permitting head-wrestling. With weapons that can be used harmlessly, sporting engagements can flourish, and they do. Called "sparring", these are much engaged in by large and small, particularly outside the rutting season.

Gregarious species that defend mating territories retain fairly simple grappling horns, but in species with serial or harem defence of females, the horns may enlarge. Huge antlers or horns may evolve, as in reindeer, the extinct Irish elk, and the water buffalo.

What artiodactyls do, no matter where they may live, is a logical consequence of generating security, minimizing competition for resources, maximizing the extraction of nutrients and energy from fibrous, toxic plants, and maximizing the chances of successful, frequent reproduction. Their lives illustrate how the same problems have been diversely and ingeniously solved.

Jeff Foott/Bruce Coleman Ltd

▲ *The distinctive patches on the rumps of many gregarious animals act as signals to other members of the herd: their message is "follow me". The patch is particularly prominent in pronghorns, shown here.*

Jonathan Scott/Planet Earth Pictures

◄ *Vast herds of wildebeests (brindled gnu) migrate annually across the plains between Kenya and South Africa, grazing as they go.*

▲ *Domestic pigs are direct descendants of the wild boar of Eurasia and North Africa. The young of the wild form have a camouflage pattern of broken white stripes. Feral descendants of European breeds of pigs revert to an appearance very like that of the wild boar shown here but their young are plain colored.*

PIGS, PECCARIES, AND HIPPOS

The pigs of the Old World (eight species) and the unrelated pig-like peccaries (three species) of the New World have evolved in similar ways and are externally similar to one another. The peccaries have more complex stomachs and fused cannon bones in the hindlegs, indicating a longer history of herbivory and of sprinting from predators. Pigs and peccaries exploit, in part, concentrated food

sources below the surface of the ground, such as roots and tubers. Pigs may also take carrion, birds' nests, newborn mammals, and small rodents.

Peccaries differ from pigs by living in closely knit groups that jointly defend territories; these are usually much smaller than the home ranges roamed over by pigs. Group size varies, but may number over one hundred in the white-lipped peccaries. As a consequence of putting priority on defense, male and female have both evolved as "fighters": they have sharp canines and are alike in external appearance and size. They also save on reproduction by bearing few and very small young. Unlike pigs, peccaries expose their newborn young to the environment and so do not inhabit cold climates. Peccaries range in head and body length from 80 to 120 centimeters (30 to 48 inches) and weigh 17 to 43 kilograms (38 to 95 pounds).

Females of wild boar shelter their tiny but numerous young against cold, snow, and rain by building domed nests and by "brooding" the young, so wild boar can give birth at any time of the year, including winter. Moreover, wild boar, freed from the need to synchronize birth with warm seasons, may bear several litters a year.

Pigs are a diverse group that evolved in Africa. Here giant species appeared early in the Ice Ages, and even today peculiar species with huge tusks, such as the wart hog or giant forest hog, exist. These large tusks are used in fighting, much like short, curved horns in bovids, as a means of defense to hold and control the opponent's head. An aberrant tropical island species with large, ornate tusks is the babirusa; another aberrant form is the pygmy hog from the eastern Himalayan foothills. Compared with the giant forest hog, which may reach 275 kilograms (610 pounds), the pygmy reaches only 10 kilograms (22 pounds).

Pigs rely for security, in part, on confronting and attacking predators. Otherwise they hide in dense vegetation. Large males tend to be solitary and a hierarchy controls access to estrous females.

Jane Burton/Bruce Coleman Ltd

Short, sharp tusks for attack are matched in dominant wild boar males by thick dermal shields on the shoulders and sides for defense. Small males tend to roam in bachelor groups. Females live in mother–daughter kinship groups within large home ranges; pigs are not known to defend territories, but they do share resources. Home ranges are marked with glandular secretions. The gregarious nature of pigs and the lack of territorial defense make them ideal for domestication. They have been domesticated in many cultures since the early Neolithic. Domestic pigs have reinvaded natural habitats as feral populations in North America, Australia, and New Zealand.

▲ With a weight of about 230 kilograms (500 pounds), the pygmy hippo is only about a tenth of the mass of the "true" hippopotamus. Inhabiting dense tropical forests, it is solitary and less aquatic than its larger relative.

◄ Hippopotamuses spend the day together in water. At sunset, each individual moves out along an established path (marked at intervals by piles of its dung) to a grazing area. They return around sunrise. A hippopotamus is rarely aggressive to humans unless it is provoked or encountered on its marked path.

Jonathan Scott/Planet Earth Pictures

Gunter Ziesler/Bruce Coleman Ltd

▲ *The vicuña, a wild camel from the high Andes of South America, is the smallest member of the camel family. Like the related llamas, it has a long neck and legs, but no hump.*

A close relative of pigs, the hippopotamus is entirely herbivorous, fermenting grasses in a large, complex stomach. The food habits of the pygmy hippo, which lives in moist, tropical forests, are not known. Like the pigs, hippos do not have cannon bones. Rather, the four weight-bearing toes each have a metatarsal or metacarpal bone. This is a primitive feature. It suggests that pigs, peccaries, and hippos evolved, at an early stage, a way of life that did not require speedy running on hard surfaces. Hippos illustrate this well: they seek security within groups in water, but wander out at night to feed on closely cropped patches of grass. An aquatic existence, or life in high humidity, is apparently dictated by a skin that offers little protection against dehydration. Grouping may protect the young against crocodiles, but the mortality of young is high.

Hippos use tusks in combat and have a thick dermal armor to protect themselves. Large males place a territory over areas with aquatic groups, but tolerate submissive males. Hippos have a long life and low birth rate.

During warm periods in the Ice Ages hippos were found as far north as England. They also colonized Mediterranean islands, where they shrank greatly in size. These island dwarfs, along with dwarf elephants and dwarf deer, fell victim to

Lee Lyon/Bruce Coleman Ltd

CAMELS

Camels, which evolved in North America, diversified in the Tertiary from tiny gazelle camels to huge giraffe-like browsers. They died out in North America during the Ice Ages, but some had emigrated and survived in South America, Eurasia, and North Africa. The two species of Old World camels, whose ability to withstand dehydration is legendary, are desert-adapted social mammals that bear large, well-developed young after long gestation periods. They also have red blood cells that contain a nucleus, an anomaly among mammals. They are able to feed on dry, thorny, desert vegetation and move long distances when they detect distant rainfall and green pastures. Both the dromedary and Bactrian camels average in height at the hump 210 centimeters (85 inches) and in weight 550 kilograms (1,200 pounds). Like their small South American cousins, the Old World camels are domesticated. The last of the wild camels in the cold deserts of northwest China and Mongolia are greatly endangered.

There are still natural populations of wild South American camels such as the guanaco and vicuña—the latter very vulnerable. The South American camels are much smaller than the Old World species and have no humps. The vicuña, the smallest of the South American species, averages 91 centimeters (36 inches) in shoulder height, and 50 kilograms (110 pounds) in weight. South American camels are actually North American in origin and have been in South America only since the early Pleistocene (about 2 to 3 million years ago). Males tend to defend a territory which is used by a group of bonded females and their young.

All camels have remarkable combat teeth, in which not only the canines but also the first premolars have been formed into caniforme fighting teeth.

▲ *There are about 15 million domesticated camels in the world. A few wild Bactrian (two-humped) camels survive in the Mongolian region and more than 25,000 dromedaries (one-humped camels) thrive in a feral state in the deserts of Australia. These latter are descendants of domesticated animals that were abandoned early this century.*

the first wave of Neolithic settlers about 8,000 years ago. The pygmy hippo of Liberia and Ivory Coast is distinctly less specialized for aquatic life than the hippo. It has relatively longer legs, less webbing between the toes, and less protrusion of the eyes to allow better vision above water. It lives apparently in family groups in moist forests and swamps, and depends on wallows much as the larger and more specialized hippo. The pygmy hippo has an average head and body length of 1.6 meters (64 inches) and weight of 230 kilograms (500 pounds), in contrast to the hippo's average length of 3.4 meters (125 inches) and weight of 2,400 kilograms (5,300 pounds).

▲ *The water chevrotain, which inhabits the rainforests of Western Africa, is about the size of a rabbit. Lacking antlers, it defends itself with sharp canine teeth. It often seeks shelter in hollow trees, within which it climbs with agility.*

MOUSE DEER AND MUSK DEER

The most primitive artiodactyls are the four species of tropical tragulids, the mouse deer of Asia, and the water chevrotain of Africa. These are the remnants of a family abundant in the early Tertiary. Though superficially similar to small deer, tragulids are in many respects closer to pigs. They have a simpler rumen than the others, are diverse in food habits, and are territorial resource

defenders armed with ever-growing, sharp, fighting canines. The premolars still have conical crowns, much like those of the primitive ungulates from the early Tertiary. Water chevrotains lack cannon bones in the front legs. As in other small tropical resource-defenders, the female may be larger than the male. Water chevrotains average 75 centimeters (30 inches) in head and body length, 35 centimeters (14 inches) in shoulder height, and 10 kilograms (20 pounds) in weight.

Close to tragulids are the three species of musk deer, which can be viewed as "improved and enlarged" cold-adapted tragulids. In fact, the Siberian musk deer penetrates further north than the large wapiti. The musk deer differs enough from tragulids and deer to warrant separate family status. It lives in temperate or cold climates within continental Asia, where it feeds on small bits of highly digestible plant matter, including shoots, lichens, fruit, and soft grasses. It runs like a rabbit in long jumps, propelled by powerful hindlegs. It is also a surprisingly good climber of rocks and large, well-branched trees. Larger than most tragulids, it rivals the roe deer in size: it averages 90 centimeters (36 inches) in head and body length, 60 centimeters (24 inches) in height, and 12 kilograms (26 pounds) in weight.

Only the males carry long combat canines. They are strongly territorial and the apple-sized musk

▼► *The Indian muntjac (below) is a small deer, weighing about 22 kilograms (48 pounds). Males have very simple antlers and well developed canine tusks. Muntjacs are either solitary or move about in pairs or small family groups. Most of the world's fallow deer (right) now live under some degree of protection in deer parks or reserves. Males may weigh more than 100 kilograms (220 pounds). The prominent Adam's apple is characteristic.*

gland, which is situated between the penis opening and the umbilicus, may serve in olfactory communication. The spotted young are very small compared to those of other deer; breeding is seasonal. Much is to be learned yet about the musk deer, whose valuable musk has made it a target for commercial exploitation. It is heavily snared and hunted, which has led to local scarcity or extinction. The musk of the male is used in the folk medicine of the Far East as well as by perfume manufacturers. Experimental projects to keep musk deer in captivity and periodically remove the musk of males are still beset by difficulties. The musk deer is just one of the many species of wildlife whose existence is threatened by the economic rewards placed on their dead bodies.

DEER

The Cervidae or deer family (39 species) consists of the Old World deer, the New World deer, and the aberrant, antlerless water deer of Korea, *Hydropotes inermis*. Deer are characterized by bone antlers that are grown and shed annually. Antlers have reached the monstrous dimensions of a 3.5 meter (12 foot) span in the extinct Irish elk and a weight of just under 50 kilograms (110 pounds). Deer are advanced ruminants that are tied to woody vegetation, though one species, the highly gregarious reindeer, has adapted to tundra. Deer never invaded sub-Saharan Africa, but they reached tropical South America just before the Ice Ages and there diversified into many small-bodied species.

▼ One of the largest of the living deer, weighing up to 450 kilograms (990 pounds), the wapiti occurs naturally in north-western North America. Unusually for an animal living seasonally in snow, its coat is darker in winter than in summer. Unlike its close relative the red deer, which roars in the rutting season, the wapiti "bugles"—a high-pitched whistling sound.

◀▼ Antlers are very different from horns. Horns are permanent structures made of the same substance as hair or claws. Antlers consist of bone and are shed and regrown each year. While they are growing, antlers have a covering of skin, called "velvet"; when growth is complete, the skin peels off or is rubbed off. There is a broad correlation between the size of a deer and the size and complexity of its antlers. The water deer has no antlers. Small species such as the pudu (below, left)—which is in fact the smallest species of deer, with a shoulder height of 35 to 38 centimeters (14 to 15 inches)—have simple, unbranched antlers. Large species such as the moose (top, left)—the largest species of deer, with a shoulder height of 168 to 230 centimeters (5 feet 7 inches to 7 feet 8 inches)—and reindeer or caribou (below, right) have enormous, many-branched antlers, often with webs of bone between the branches. The reindeer is the only species of deer in which both sexes have antlers.

WHY ARE ANTLERS SO LARGE?

This question was addressed by Charles Darwin. Quite correctly he saw here a parallel to the showy feathers of pheasants and peacocks. Darwin suspected that these organs somehow evolved by female choice, but he did not resolve how. The mystery surrounding large horns is lifting, but the explanation at first sight borders on the incredible: large horns and antlers are an indirect consequence of the security adaptations of newborns.

In a species adapted for running, the young must soon run as fast and with as much staying power as the mother. The dangerous post-birth period can be shortened by making the young as large and as highly developed at birth as possible, and by supplying it with much or rich milk so that it grows rapidly, and this adaptation has occurred.

A mother following these rules must be very good at saving nutrients and energy from her maintenance costs and body growth towards the growth and feeding of the young. But what about father? If he spares nutrients from growth and maintenance they at once increase horn or antler growth. That, however, should be an advantage, because now the male carries a symbol of his biological success on his head. Females, then, should choose males with big antlers to ensure large neonates for and copious milk production by their daughters. After all, the bigger the antlers, the better

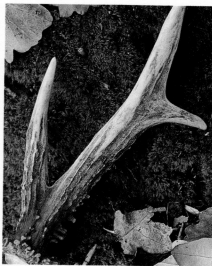

Hans Reinhard/Bruce Coleman Ltd

▲ *Shed antler of a roe deer.*

▼ *Male red deer with antlers in velvet.*

the male was at obtaining surplus resources, and the more efficient he was at maintenance. In addition, symmetry of antlers is an excellent indicator of health. We therefore expect that in species with larger horns or antlers the males will be more likely to show off with them in courtship, but not in dominance displays to other males.

These predictions work out: the antler size of the father correlates with the size of the young at birth, and with the percentage of solids in the milk of female descendants. The larger the antlers, the more they are shown to females in courtship, but not to males in display. In small-antlered animals, where female choice is expected to be minor, we expect males to gain access to females by violence. In that case males are better at shunting their surplus energy and nutrients into body growth as body size is vital to victory. That is, if males are small-horned and not territorial, we expect them to grow larger in body size year by year; in large-antlered species we expect males to plateau soon in body size. That also appears to be the case. Not surprisingly, the extinct Irish elk, with its huge antlers, turns out to be the most highly evolved runner among deer. And the reindeer, which has the largest antlers and is the most highly evolved runner among living deer, has highly developed, relatively large young and the richest milk among deer.

Hans Reinhard/Bruce Coleman Ltd

Masood Qureshi/Bruce Coleman Ltd

▲ Because they feed upon the leaves of trees, giraffes have a year-round food supply and therefore can breed at any time, unlike other hoofed animals of the African plains which must adjust their breeding to seasonal variation in the abundance of ground vegetation.

GIRAFFES

The family Giraffidae is now reduced to two species: the tropical, forest-adapted okapi, a species similar to primitive giraffes of the mid-Tertiary; and the largest ruminant of all, the giraffe of the open African plains. Giraffe species were more numerous in Tertiary times. They also evolved several lineages of grazing giraffes with ox-size bodies and large "horns", such as the sivatheres. In giraffes, as in bovids, the horns are formed from ossicones, but giraffes grow a covering of hair, not horn, from the skin over the ossicones. The ossicones continue to get a cover of bone throughout life, so that a bull's skull grows increasingly massive with age. In both species of giraffe the skull is used like a club in fighting and the skin is enlarged into a thick dermal armor to counter blows from the mace-like head.

Giraffes also share a unique courtship behavior with bovids, suggesting a common descent. The giraffe is a highly specialized, very successful foliage feeder which, by obtaining green foliage year round, has escaped the limitation experienced by ground-level grazers in the savanna: a seasonal supply of easily digested green vegetation. It can thus reproduce year-round. Giraffes, like many plains ungulates from productive landscapes, give birth to very large, highly developed young. Being freed from the constraints of a seasonal food supply, giraffes can also grow very large. Male giraffes have an average head and body length of 4.2 meters (14 feet), an average height to the horn tips of 5 meters (16 feet), and an average weight of 1,350 kilograms (3,000 pounds). Females average 4.2 meters (14 feet) in height and 870 kilograms (1,900 pounds) in weight.

Giraffes mature relatively slowly, bear few young, but have a long potential life-span. Bulls gain access to cows by means of a dominance hierarchy. Despite their odd body shape, giraffes are good runners.

The okapi, a forest giraffe, is also a foliage feeder, and in its biology appears to share many similarities with its advanced relative from the open plains. It has an average head and body length of about 2 meters (7 feet), an average height of 36 centimeters (15 inches), and an average weight of about 230 kilograms (500 pounds). Today it is highly localized geographically and, though protected, shares the scourge of most species of large mammals: illegal killing for commerce in meat and parts.

PRONGHORNS

The pronghorn is a gazelle-like ruminant, the only species of the many peculiar indigenous North American ruminants to survive to the present day. Like other large North American mammals to survive the great extinction at the end of the Ice Ages, pronghorns are ecological opportunists and have high reproductive rates. These strikingly colored, keen-eyed "antelopes" live gregariously in open plains as speedy, enduring runners; they are very light in build and deposit little fat on their bodies. Their average weight is 60 kilograms (130 pounds); their head and tail length averages 140 centimeters (55 inches); and shoulder height 87 centimeters (34 inches). By forming breeding territories in summer, pronghorns have not only a long gestation period, but also a breeding system surprisingly similar to that of the forest-adapted roe deer.

Pronghorn bucks also share social signals with gazelles. The branched horn sheath regrows annually, but from the tips of the horn cores only. Females may also carry short horns. The horn sheaths are shed annually right after the mating season. Female pronghorns superovulate so that a handful of embryos implant in the uterus. Here the embryos kill one another by growing long outgrowths through the bodies of rival fetuses until only two survive to grow to term. Like other gregarious plains runners, pronghorns have large young. They mature rapidly, but have a short life-span as adults. Twin births are the norm.

▼ The okapi (below, left) was not known to Europeans until 1900. It is a short-necked giraffe which browses in equatorial rainforest. Its color pattern provides excellent camouflage in its natural environment. The pronghorn (below, right), the only member of the Antilocapridae family, lives in North America. Often referred to as an antelope, it is not really a member of that group. Pronghorns are the fastest-running mammals in North America.

Hans Reinhard/Bruce Coleman Ltd

▲ *Thick, shaggy hair and a densely matted undercoat provide the yak with insulation against the cold of the Himalaya mountains.*

THE BOVID FAMILY

The largest group of artiodactyls, with about 107 species, is the family Bovidae. It includes cattle; goat-antelopes, shrub and musk oxen, true goats and sheep; gazelles and their relatives; the primitive duikers; twisted-horned antelopes, such as the eland and kudu; reed and water bucks; roan and sable antelopes; gnus and hartebeests; the dwarf antelope; plus several odd species such as the four-horned antelope and the blue bull of India.

Bovids have an Old World origin and are currently distributed from hot deserts and tropical forests to the polar deserts of Greenland and the alpine regions of Tibet. No indigenous bovid species are found in South America or Australia. They are characterized by non-deciduous horns that grow from an ossicone on the forehead. This ossicone forms in the skin of the forehead and attaches to the frontal bone. It is largely hollow inside, giving rise to the designation of bovids as hollow-horned ruminants. The diversity of horn shape and size is striking, as is the diversity of ecological adaptations and body sizes.

Bovids form the bulk of the tropical grazers and desert dwellers and supply most livestock species such as cattle, sheep, and goats. Extinctions at the end of the last Ice Age affected mainly large-bodied species in Africa, Europe, and North America; most of the smaller species survived to the present. Domestication of sheep and goats began early in the Neolithic era. The small wild sheep of Corsica, Sardinia, and Cyprus, as well as the wild goats on Crete, appear to be non-native species derived from transplants by Neolithic people. On some islands feral goats have gravely damaged indigenous vegetation and associated animals. Species of interest to hunters have also been widely distributed by human hand. Such "exotics" are now a threat to native wildlife in North America.

VALERIUS GEIST

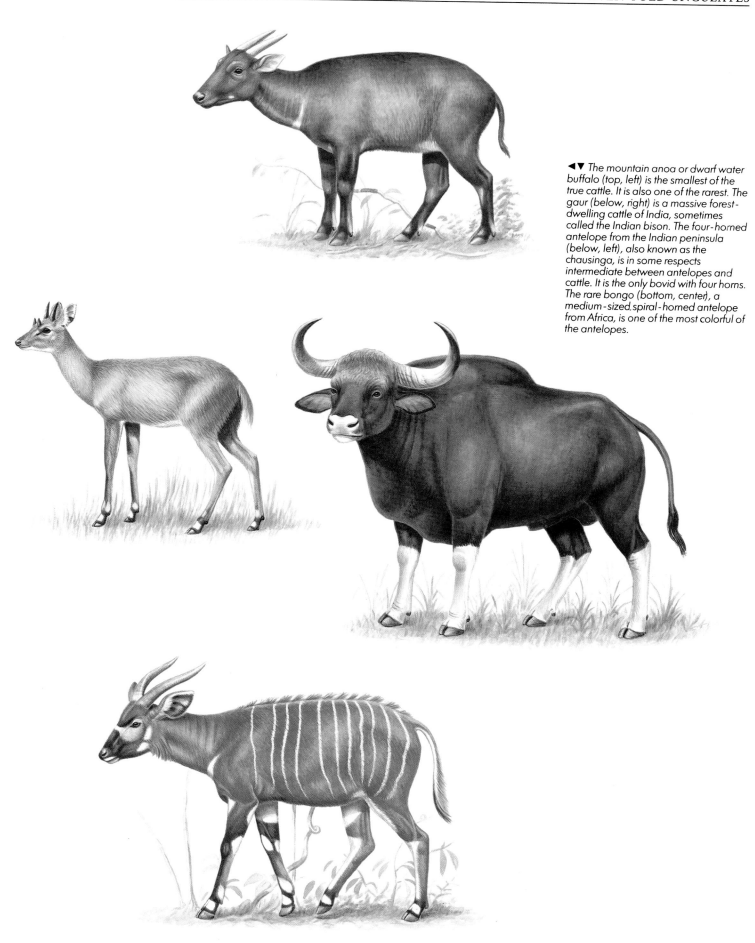

▲▼ The mountain anoa or dwarf water buffalo (top, left) is the smallest of the true cattle. It is also one of the rarest. The gaur (below, right) is a massive forest-dwelling cattle of India, sometimes called the Indian bison. The four-horned antelope from the Indian peninsula (below, left), also known as the chausinga, is in some respects intermediate between antelopes and cattle. It is the only bovid with four horns. The rare bongo (bottom, center), a medium-sized spiral-horned antelope from Africa, is one of the most colorful of the antelopes.

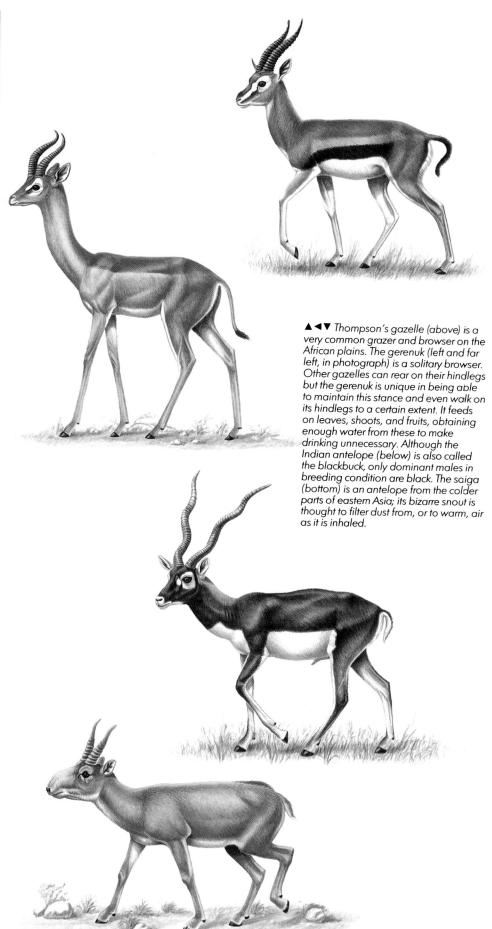

▲ ◄ ▼ Thompson's gazelle (above) is a very common grazer and browser on the African plains. The gerenuk (left and far left, in photograph) is a solitary browser. Other gazelles can rear on their hindlegs but the gerenuk is unique in being able to maintain this stance and even walk on its hindlegs to a certain extent. It feeds on leaves, shoots, and fruits, obtaining enough water from these to make drinking unnecessary. Although the Indian antelope (below) is also called the blackbuck, only dominant males in breeding condition are black. The saiga (bottom) is an antelope from the colder parts of eastern Asia; its bizarre snout is thought to filter dust from, or to warm, air as it is inhaled.

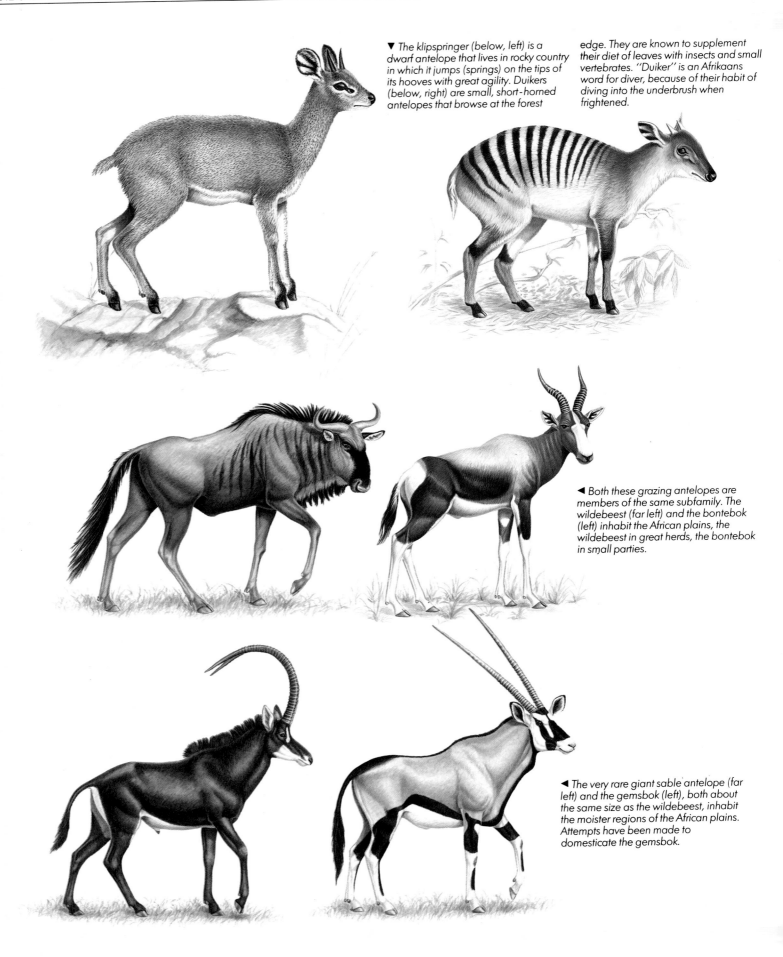

▼ The klipspringer (below, left) is a dwarf antelope that lives in rocky country in which it jumps (springs) on the tips of its hooves with great agility. Duikers (below, right) are small, short-horned antelopes that browse at the forest edge. They are known to supplement their diet of leaves with insects and small vertebrates. "Duiker" is an Afrikaans word for diver, because of their habit of diving into the underbrush when frightened.

◄ Both these grazing antelopes are members of the same subfamily. The wildebeest (far left) and the bontebok (left) inhabit the African plains, the wildebeest in great herds, the bontebok in small parties.

◄ The very rare giant sable antelope (far left) and the gemsbok (left), both about the same size as the wildebeest, inhabit the moister regions of the African plains. Attempts have been made to domesticate the gemsbok.

◄▼ All the goats and sheep illustrated here are males that compete by head-butting. The wild goat of Western Asia (top, left) is the ancester of all the domestic breeds of goat. The Himalayan tahr (below, left) is a long-haired caprine. Bighorn sheep (top, right) live in North America, where they are found in a wide variety of habitats ranging from cold alpine areas to deserts. The Barbary sheep, also known as the aoudad (below, right), lives in northern Africa. Despite its size and appearance, the musk ox (bottom) is more closely related to goats than to cattle. The horns are adapted to competition by head-butting and are also used in defense.

PANGOLINS

▼ *Pangolins are the only mammals to have a covering of scales. These are made of horn.*

At one time classified together with the anteaters, sloths, and armadillos in the order Edentata, the pangolins, or scaly anteaters, have now been placed in their own distinctive order, the Pholidota. The pangolins are a single family, Manidae, and are represented today by two genera, the Asian pangolins, *Manis,* and the African pangolins, *Phataginus.* Distinguishable from all other Old World mammals by their unique covering of horny body scales, pangolins are truly strange-looking animals that, at a glance, appear more reptilian than mammalian. These body scales, which grow from the thick underlying skin, protect every part of the body except the underside and inner surfaces of the limbs. With a tail that, in some species, is twice the length of the entire body, small limbs, a conical head, and short but powerful limbs, this is one of the most extraordinary-looking mammals of the Old World tropics.

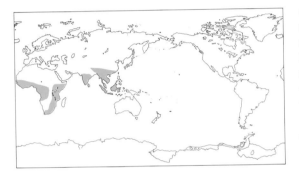

MAMMALS WITH SCALES

Pangolins occur in much of Southeast Asia and in tropical and subtropical parts of Africa. The three Asian species possess external ears, have a scale-clad tail, and also have hairs at the base of the body scales. The four African species do not have external ears; the rear part of the breastbone is very long; and the tail has no scales beneath. Two species, the giant (*Phataginus gigantea*) and the Cape (*P. temmincki*) pangolin, are terrestrial, while the long-tailed (*P. tetradactyla*) and small-scaled (*P. tricuspis*) tree pangolins are, as their names suggest, arboreal. The giant pangolin has a body length of 80 to 90 centimeters (31 to 35 inches) and a tail measuring 65 to 80 centimeters (26 to 31 inches); it weighs 25 to 35 kilograms (56 to 78 pounds). The body of the long-tailed pangolin ranges from 30 to 35 centimeters (12 to 14 inches) in length, the tail, from 50 to 60 centimeters (20 to 24 inches); it weighs 1.2 to 2 kilograms (2½ to 4½

pounds). In all species, the male is much larger than the female—the male Indian pangolin may be as much as 90 percent heavier than the female.

All species have short but powerful limbs, which are used for digging into termite mounds and anthills. The terrestrial species also use their claws for scooping out underground burrows where they conceal themselves during the day. The arboreal species seek refuge in tree hollows, curling themselves up for protection when asleep. The tail, though covered with scales, is fairly mobile, and, in some forms, even prehensile. In addition, the tail has two other important functions: it is very sensitive at the tip and, even when not prehensile, can be hooked like a finger over a solid support. If threatened, a pangolin can also lash its tail at an adversary, using the razor-sharp scales with devastating effect. This action may also be supplemented by spraying an attacking animal with a foul-smelling fluid from the anal glands.

Pangolins feed exclusively on insects—basically termites and ants—catching them with their long, proboscis-like tongue. Housed in a special sheath attached to the pelvis, the 70 centimeter (27 inch) tongue is coiled up in the animal's mouth when at rest. Viscous saliva secreted onto the tongue by special glands in the abdomen traps its prey when the tongue is flicked into the chambers of the mound. Pangolins have no teeth and all food is crushed in the lower section of the stomach leading to the intestines. This region usually contains small pebbles and seems to function by grinding food in the same manner as the gizzard of a bird. When it is feeding, thickened membranes protect the pangolin's eyes, and special muscles seal its nostrils to shield it from the bites of ants.

The degree of sensory development among these different species is directly related to the animal's diet and way of life. Largely nocturnal, pangolins have a poor sense of vision and only average hearing. The sense of smell, however, is exceptionally acute, and this probably plays a major role in communication. Pangolins are largely solitary animals that do not appear to actively defend a fixed territory from neighboring animals of the same species. However, by repeatedly marking selected trees and rocks with secretions from the anal gland, a pangolin notifies neighboring and potentially intruding animals that the area is already occupied.

CONSERVATION

In Africa, large numbers of pangolins are killed each year for meat, while in Asia the Chinese have traditionally attributed medicinal values to the scales and for that reason the animals have always been relentlessly hunted. Pangolins are unlikely to be replaced by more adaptable competitors, but the greatest single threat to the survival of these strange-looking creatures is the destruction of their habitats, which will certainly

M.P.L. Fogden/Bruce Coleman Ltd

have a severe effect on animals with such highly specialized, restricted feeding behavior. Formerly extensive tracts of tropical rainforests are being severely eroded and irreparably damaged each day. Without human action to protect what remains of their habitats, these species are unlikely to survive.

R. DAVID STONE

▲ Although the Chinese pangolin climbs with agility it feeds mainly on the ground, digging for termites with its strongly clawed feet.

▼ The tree pangolins of Africa climb with the aid of a very prehensile tail; the single young uses its tail to cling to its mother.

Keith & Liv Laidler/Ardea London

RODENTS

ORDER RODENTIA
• 29 families • 443 genera
• c. 2,000 species

SMALLEST & LARGEST

Pygmy jerboa *Salpingotulus michaelis*
Head–body length: 36–47 mm
(1²⁄₅–1⁹⁄₁₀ in)
Tail length: 72–94 mm
(2⁴⁄₅–3⁷⁄₁₀ in)
Weight: 5–7 g (c. ¹⁄₅ oz)

Capybara *Hydrochoerus hydrochaeris*
Head–body length: 106–134 cm
(42–53 in)
Weight: 35–64 kg (77–141 lb)

CONSERVATION WATCH
!!! 66 species are listed as critically endangered, including: giant kangaroo rat *Dipodomys ingens*; Mt Isarog striped rat *Chrotomys gonzalesi*; 13 gerbil species (genus *Gerbillus*); western small-toothed rat *Macruromys elegans*; short-tailed chinchilla *Chinchilla brevicaudata*; large-eared hutia *Mesocapromys auritus*; Cabrera's hutia *Mesocapromys nanus*; little earth hutia *Mesocapromys sanfelipensis*; Garrido's hutia *Mysateles garridoi*.
!! 98 species are endangered.
! 166 species are listed as vulnerable.

The success of rodents can hardly be in doubt. They make up just under 40 percent of all mammal species and have a worldwide distribution. They have adapted to habitats from the high Arctic tundra to tropical deserts, forests, and high mountains. Rodents have also reached—often with human help—some of the most isolated oceanic islands. Their relationships with humans are often close and frequently deleterious: historically, they have spread fatal diseases on an enormous scale, and they consume crops and stored products.

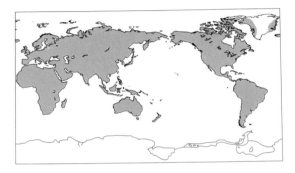

INGREDIENTS OF SUCCESS

To what do these animals owe their great success? Three main factors come to mind. First, although there was major evolutionary diversification during the Eocene (about 57 to 37 million years ago), one family, the Muridae, did not appear until the Pliocene (about 5 million years ago) and is therefore relatively young. With more than 1,000 species, this family is still maximizing its genetic diversity. Throughout its evolution this group has remained relatively unspecialized.

Second, rodents are generally small: most weigh less than 150 grams (5 ounces), though there are many exceptions, with the capybara weighing up to 66 kilograms (145 pounds). Small size affords a good opportunity for the exploitation of a wide range of microhabitats.

Third, many rodents are reproductively prolific: short gestation periods, large litters, and frequent breeding are characteristic, permitting survival under adverse conditions and rapid exploitation under favorable ones.

The combination of evolutionary flexibility, small size, and high production have permitted relatively modest structural and functional modifications to be sufficient to produce the diverse array of contemporary species.

ADAPTATION FROM A SIMPLE PLAN

Rodents have a remarkably uniform mouse-like body plan that has been subject to modification principally of the teeth and digestive system, the limbs, and the tail. All rodents have a single pair of open-rooted, continuously growing sharp incisors to gnaw into food. Behind these is a space or diastema whose presence permits the lips to be brought together to exclude unwanted particles of gnawed material. There are no canines. At the back of the mouth is a row of molar, and sometimes premolar, teeth, usually used for grinding food. Molars and premolars vary in number from 4 in the one-toothed shrew mouse *Mayermys ellermani* to 24 in the silvery mole rat *Heliophobius argentocinereus*; there are commonly 12 or 16. The chewing surface may consist of ridges or cusps.

Although most rodents are herbivorous and have a relatively large cecum to house the bacterial flora used in cellulose digestion, many have alternative feeding habits and appropriately adapted digestive systems. Examples include omnivory in the house rat *Rattus norvegicus* and the Arabian spiny mouse *Acomys dimidiatus*; insectivory in the speckled harsh-furred rat *Lophuromys flavopunctatus*; and carnivory in South American fish-eating rats of the genera *Anotomys* and *Ichthyomys* and the Australian water rat *Hydromys chrysogaster*.

Most rodents walk on the soles of their feet. In jumpers, such as the jumping mice and jerboas, the hindfoot is greatly elongated and only the toes reach the ground surface. Climbers have opposable big toes (for example, the palm mouse *Vandeleuria oleracea*), hands or feet, or both, broadened to produce a firmer grip (for example, Peter's arboreal forest rat *Thamnomys rutilans*), or sharp claws (tree squirrels). For rapid running, the agouti *Agouti paca* has elongated limbs with only the fingers and toes reaching the ground. Aquatic forms can have long and slightly splayed hindfeet, as in the African swamp rats of the genus *Malacomys*, or webbing, as in the Australian water rat.

The bushy tails of squirrels serve for balance and, in the African ground squirrel *Xerus erythropus*, when recurved over the body, for shade; aquatic forms may have a horizontally flattened tail for swimming (as in the beaver, *Castor* species), a laterally flattened tail to act as a rudder (muskrat *Ondatra zibethicus*), or a longitudinal fringe of hairs running under the tail and adding to its surface area (earless water rat *Crossomys moncktoni*). The rapid runners and jumpers commonly have a long balancing tail, often with a distinct tuft of hairs at its tip, as in jerboas. In the harvest mouse *Micromys minutus* the tail is a grasping organ used in climbing. The tail may also be used for communication. The elaborate fan of Speke's gundi *Pectinator spekei* is for both balance and social display. The smooth-tailed giant rat *Mallomys rothschildi* has a dark tail with a white end which probably has a behavioral function.

The three major groups of rodents are separated on the basis of their jaw musculature. The muscles involved are the deep and lateral branches of the masseter which pull the lower jaw forward (in gnawing) and close the lower on the upper jaw. Their functions vary in different groups, depending on the position of the muscle branches. In the Sciuromorpha or squirrel-like rodents the lateral masseter brings the jaw forward and the deep masseter closes the jaw; in the Caviomorpha or cavy-like rodents the lateral masseter closes the jaw and the deep masseter provides the gnawing movement; and in the Myomorpha or mouse-like rodents both branches of the masseter are involved in gnawing. The last provides the most effective gnawing action. The parts of the skull associated with these different muscle functions are quite distinct for each of the three groups.

SQUIRREL-LIKE RODENTS

Apart from the sciuromorph jaw musculature, the seven families of this group have little in common and probably diverged early in rodent evolution. Squirrels account for almost 74 percent of the species. With a worldwide distribution, except Australia, Polynesia, southern South America, and the Sahara and Arabian deserts, squirrel-like rodents are to be found in most habitats. They range in size from the pocket mouse *Perognathus flavus*, which weighs 10 grams (⅓ ounce) to the beaver *Castor canadensis*, which weighs 66 kilograms (145 pounds).

Mountain beaver and beavers

The mountain beaver *Aplodontia rufa* is the most primitive of living rodents. It weighs 1 to 1.5 kilograms (2 to 3 pounds) and lives in coniferous forests in North America, where it constructs an elaborate burrow inhabited for much of the year by one animal. The mountain beaver has difficulty regulating its temperature and conserving body moisture, which makes hibernation and summer torpor impracticable.

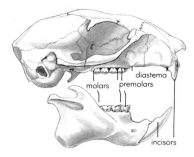

▲ The skull of a rodent can be recognized by a pair of continuously growing, chisel-edged incisors in the upper and lower jaws, and a long gap between these and the grinding teeth.

RODENT SKULL MUSCULATURE

SCIUROMORPHA

MYOMORPHA

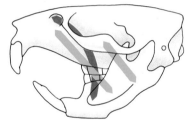

CAVIOMORPHA

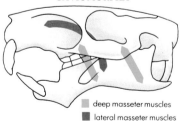

▊ deep masseter muscles
▊ lateral masseter muscles

▲ The major jaw muscles (masseters) are arranged differently in the three suborders of rodents. In the diagrams above, the orientation of the lateral masseter is shown in red and that of the deep masseter in blue. In the squirrel-like rodents (top), the lateral masseter pulls the lower jaw forward and the deep masseter closes the jaws. In the mouse-like rodents (center), the lateral and deep masseters work together to pull the jaw forward. In the cavy-like rodents (bottom), the lateral masseter closes the jaws and the deep masseter pulls the jaw forward.

◀ Rodents are specialized as gnawing animals, none less so than beavers, which fell substantial trees with their teeth. Pictured is the European beaver.

Hans Reinhard/Bruce Coleman Ltd

The North American beaver *Castor canadensis* and European beaver *Castor fiber* are large herbivores, highly adapted to a semi-aquatic life with streamlined body, flattened tail, and webbed feet. They live in closed, hierarchical family units consisting of an adult pair and the offspring of up to several previous years; they have one litter each year. Using their incisor teeth, beavers cut down trees for food and for building dams across streams to impound water and create ponds. They build conical lodges in the ponds with access to the living chamber through an underwater tunnel.

A selection of arboreal squirrels. Upper left, the Eurasian red squirrel; right, the black giant squirrel; and bottom left, the southern or least flying squirrel. The tail is usually long in arboreal squirrels and is an important aid to balance.

Squirrels

Squirrels are a fairly uniform, little-specialized group that have successfully adapted to many habitats from rocky cliffs and semi-arid deserts to temperate and tropical grasslands and forests, in all continents except Australia. They can be divided into diurnal arboreal species, nocturnal arboreal gliders, and diurnal ground-dwellers. They range in size from the African pygmy squirrel of about 10 grams (⅓ ounce) to the Alpine marmot of up to 8 kilograms (17½ pounds).

It is probably the bushy, spectacular tail, active habit, bright eyes, and general alertness that people find endearing about squirrels. With their excellent eyesight and wide vision, tree squirrels can precisely appreciate distance in three dimensions. Movement through the trees is facilitated by sharp claws on the digits and a counterbalancing tail. When the animal jumps the limbs are outstretched, the body flattened, and the tail slightly curved, presenting as broad a surface as possible to the atmosphere. On the ground, movement is by a series of arched leaps.

Flying squirrels glide rather than fly, using a membrane down each side of the body, called the patagium, as a parachute and the tail as a rudder.

Direction is controlled by the legs, tail, and stiffness of the patagium. On landing, flying squirrels brake by flexing the body and tail upward. They are less agile climbers than tree squirrels.

Ground squirrels are widespread and include prairie dogs of the genus *Cynomys* in America, marmots of the genus *Marmota,* and sousliks of the genus *Spermophilus* across the Northern Hemisphere, and the African ground squirrels of the genus *Xerus.* Many inhabit burrows where young are reared, food stored, and protection provided from predators. Many diggers have strong forelimbs and long claws. In some species, for example *Cynomys* and *Marmota,* the tail is much reduced. Many temperate species hibernate.

Most arboreal squirrels are herbivores and feed on fruit, nuts, seeds, shoots, and leaves, though this diet can be supplemented by insects. Terrestrial species often consume grasses and herbs in the immediate vicinity of their burrows. Squirrel feeding habits can affect the environment: red squirrels damage forestry plantations by eating young conifer shoots, and bark stripping by grey squirrels *Sciurus carolinensis* destroys some deciduous trees. On the other hand, failure to recover cached acorns serves as a dispersal mechanism, and the preferential feeding of prairie dogs on herbs favors growth of grama grass.

Social organization is most highly developed in terrestrial species. Prairie dogs live in social units (coteries) consisting of an adult male, several adult females, and associated young. No dominance exists in the coterie, resources being shared within the territory. Considerable aggression is displayed between adjacent coteries, although there is some seasonal relaxation in spring and summer. Numbers of coteries with interconnecting burrow systems form extensive "towns". New territories are established when adult animals move out of the coterie. Cooperative alarms are raised at danger.

Marmots have a similar social structure, but the mountain-dwelling species, experiencing a shorter growing season, have the highest degree of sociality. Although grey squirrel home ranges may overlap, dominance hierarchies occur in both sexes, and low-status animals are forced to emigrate at times of food shortage. This system may be widespread among arboreal species as it ensures optimal use of resources, minimal competition, and social contact with other animals of the same species.

Francois Gohier/Auscape International

▲ Most ground squirrels live in small social groups, sharing a burrow system. The social unit of prairie dogs includes a male, several females, and their young. No adult is dominant over the others.

A selection of ground squirrels. Left, the Columbian ground squirrel; top center, the hoary marmot; right, the least chipmunk. The tail is relatively short in these burrowing rodents.

Pocket gophers and pocket mice

Both of these families have "pockets" or cheek pouches on either side of the mouth for holding and carrying food.

The pocket gophers, which weigh 45 to 400 grams (1½ to 14 ounces), occur in various habitats in North and Central America. They are highly adapted burrowers with tubular bodies and short powerful limbs; the hand may be broad with strong nails, as in the large pocket gopher *Orthogeomys grandis*. The incisors assist the forelimbs in burrowing. Pocket gophers eat surface vegetation and underground roots and tubers; in the process they can be serious agricultural pests.

The pocket mice are granivores of arid to wet habitats of North, Central, and northern South America. Up to six species have been recorded from the same habitat in arid regions, whereas far fewer are found in tropical rainforest. This is attributed to seeds having better chemical protection in rainforest. Most species are quadrupedal and range close to their burrows. The bipedal kangaroo rats of the genus *Dipodomys* and kangaroo mice of the genus *Microdipodops* extend over wider areas, being partially protected from predators by very efficient hearing.

Scaly-tailed squirrels

These African rainforest dwellers are, with the

▶ *European harvest mice are tiny seed-eating rodents that climb in grasses and shrubs, aided by a very prehensile tail. American harvest mice, which have a similar way of life, belong to a different subfamily.*

▼ *The spring hare is a very large burrowing rodent which bounds on its hind legs like the much smaller jerboas and hopping mice.*

G. Cubitt/Bruce Coleman Ltd

exception of one species (*Zenkerella insignis*), excellent gliders. They have a well-developed patagium and, beneath the basal region of the tail, triangular scales with outwardly projecting points, which possibly serve to help the animals to grip branches. The two pygmy species, *Idiurus macrotis* and *I. zenkeri*, weigh about 20 grams (¾ ounce) and live in colonies of up to a hundred in tree cavities. These and the larger species are herbivorous, eating fruits, nuts, leaves, and flowers. They are all probably mainly nocturnal.

The jumping hare of the African veldt

The spring hare or springhaas *Pedetes capensis* inhabits dry savanna and semi-desert in eastern and southern Africa. A nocturnal grazer, it spends the day underground in a complex burrow system with several entrances. This large rodent, which weighs 3 to 4 kilograms (6½ to 9 pounds) and has an upright posture, is an excellent jumper, having long hindlegs. The female produces one young about three times a year.

MOUSE-LIKE RODENTS

Occurring throughout the world, the mouse-like rodents or Myomorpha, which account for over one-quarter of all mammal species, have successfully adapted to all habitats except snowy wastelands, notably in Antarctica. They are generally small, the largest species being the African maned rat *Lophiomys imhausii*, which weighs 2.5 kilograms (5½ pounds).

Rats and mice

The vast assemblage that constitutes the family Muridae accommodates all but 59 of the myomorph species.

The New and Old World rats and mice belong to two distinct groups, the Hesperomyinae and Murinae, respectively. They evolved separately but have come to occupy comparable niches in their two land masses. Very closely related to the murines are the Australian water rats, Hydromyinae; they range from Australia to New Guinea and the Philippines. All three groups arose from early Oligocene (about 37 million years ago) hamster-like stocks inhabiting North America, Europe, and Asia. The establishment of a land bridge between North and South America during the Pliocene (5 to 2 million years ago) gave the ancestral hesperomyines their greatest opportunity as they entered and exploited new environments in South America. For the murines Southeast Asia was an important center for evolution. They probably evolved from an ancestral stock arriving there during the Miocene (24 to 5 million years ago). They then diversified and spread westward and eastward to establish secondary evolutionary centers in Africa and Australia. Thus in both the Old and New Worlds most speciation has taken place relatively recently, in the Pliocene and

Pleistocene (5 million to 10,000 years ago).

The absence of insectivores and lagomorphs from much of South America provided a major opportunity for the ancestral invaders to radiate. That this was achieved is illustrated by the fact that only 46 of 359 living Hesperomyinae occur in North America. In Central and South America their diversification includes omnivorous field-mice (*Akodon*); insect-eating, burrowing mice (*Oxymycterus*); arboreal fruit-eating and seed-eating climbing rats (*Tylomys*); and grain-eating pygmy mice (*Scotinomys*). There are also fish-eating rats (*Daptomys, Ichthyomys*), mollusk and fish-eating water mice (*Rheomys*), and riverbank-dwelling water rats (*Kunsia*). Numerous high altitude species occur, including the herbivorous paramo rats (*Thomasomys*), forest mice (*Aepomys*), the diurnal leaf-eared mice (*Phyllotis*), and several rice rats (*Oryzomys*).

Like the New World rats and mice, those of the Old World have much higher species densities in the tropics. For example, in Europe there are only 9 species of murine, whereas Zaire, astride the Equator, has 41 species. Even so, certain temperate genera, such as *Apodemus*, are extremely successful, as is its North American counterpart, *Peromyscus*.

The murines can be divided into two major geographical zones, Africa and Indo-Australia. Both areas are rich in species with only one genus (*Mus*) common to both. They are separated by the arid Saharo-Sindian zone. This points to relatively recent, independent evolution in the two areas.

African murines are generally small, with the African swamp rats, *Malacomys,* among the largest. These animals have a head and body length reaching about 18 centimeters (7 inches) and a weight of 130 grams (4½ ounces). In contrast, in Southeast Asia there are 10 species in the Philippines with head and body lengths over 20 centimeters (8 inches) and 6 species in New Guinea of more than 30 centimeters (12 inches). The largest representative is Cuming's slender-tailed cloud rat *Phloeomys cumingi,* over 40 centimeters long (16 inches), in the Philippines.

Three species of murine, the roof rat *Rattus rattus,* from India–Burma, the brown rat *Rattus norvegicus,* from temperate Asia, and the house mouse *Mus musculus,* from southeastern Soviet Union, have, through their close and successful association with human beings, become widespread throughout the world. In the process, the cost to humans in health, damage, and destruction has been prodigious.

Rats and mice are renowned as prolific breeders: many have gestation periods of 20 to 30 days, repeated breeding, and litters of 3 to 7. Typically, the young are born without fur, with closed eyes, and are relatively inactive. Breeding often takes place within a few months of birth. But not all conform to these characteristics. The spiny mouse *Acomys,* for example, has a gestation period of 36

NHPA/Australasian Nature Transparencies

to 40 days and usually gives birth to one or two large well-haired young. These have open eyes and are mobile within a few hours. The multimammate rat *Praomys natalensis* has an average litter size of 12, a gestation period of 23 days, and reproduction starting at 55 days; it has been estimated that one pair and their progeny could produce over 6,700 animals in 8 months.

Many murines cause havoc to agricultural crops and stored products. Among the most striking examples are the depredations of the introduced house mouse in Australia. Here populations build up to unbelievable levels: one farmer recorded 28,000 dead mice on his veranda after one night's poisoning and 70,000 killed in a wheat yard in an afternoon. In the Pacific basin and Southeast Asia three rats (*Rattus exulans, R. rattus* and *R. norvegicus*) do considerable damage to sugar cane (particularly in Hawaii), coconuts, oil palm, and rice. In Kenya, a 34 percent loss of wheat and 23 percent loss of barley crops has been reported. The species responsible were the multimammate rat, the Nile rat *Arvicanthis niloticus,* and the four-striped grass-mouse *Rhabdomys pumilio.* The first two have also seriously damaged crops of cotton in the Sudan and maize in Tanzania. Extensive

damage by these pests is not necessarily annual, but is often irregular, and, in some cases, several years may elapse between outbreaks.

Voles and lemmings

Voles and lemmings are north temperate rodents belonging to the subfamily Microtinae. They are attractive animals characterized by a blunt snout, small ears, small eyes, and a short tail. The molar teeth have flattened crowns and prisms of dentine surrounded by enamel and are highly adapted for grinding the toughest vegetation, including grasses, sedges, mosses, and herbs. Fruits, roots, and bulbs are also eaten, and seasonal changes in diet reflect availability. Food is often in shortest supply in spring and early summer.

Most of the 121 species are small, weighing less than 100 grams (3½ ounces); an exception is the semi-aquatic muskrat *Ondatra zibethicus* which can weigh 1.4 kilograms (3 pounds). With very few exceptions, they are surface dwellers and burrowers occurring in tundra, grasslands, scrub, and open forest. Some species have periodic cyclical fluctuations in numbers, sometimes attaining extremely high densities. They are active throughout the winter, surviving beneath the snow

RODENTS & HUMAN HEALTH

▼ *The black, or ship, rat appears to have originated in western Asia and to have reached Europe in the thirteenth century. It has spread to most parts of the world on ships. A closely related species from eastern Asia is almost identical in appearance and habits.*

John Markham/Bruce Coleman Ltd

Many species of rodents carry microorganisms harmful to human health. These include bacteria, rickettsias, viruses, fungi, and protozoa. Between them, these microorganisms cause many diseases. Transmission can be through an intermediate organism (vector) or direct.

Some rodent-borne diseases have been known for many hundreds of years to cause devastating mortality in human populations. One such disease is plague, which is known to be carried by over 200 species of rodent. Transmission to humans is mainly through fleas. A particularly serious threat is posed by those rodents that live near humans: the black rat *Rattus rattus* is notorious. In Central Asia two gerbil species are carriers and in southern Africa a complex transmission route exists through the main reservoir species, a gerbil, and the more commensal multimammate rat. Outbreaks of the plague are often irregular or cyclic, and are currently known to occur in Africa, North and South America, and much of temperate Asia.

Tularemia is another naturally occurring widespread bacterial disease of rodents. In Central Asia and eastern Europe voles are the main reservoirs. Transmission to humans is by various routes, including blood-sucking arthropods (mites, ticks, fleas, etc.), contaminated water, and inhaled dust containing fecal material.

The Lassa fever virus of West Africa occurs in the multimammate rat and is transmitted by human ingestion of its excreta. Similarly, the Bolivian hemorrhagic fever virus is deposited in the urine of the mouse *Calomys callosus,* a common house species in parts of South America.

For a number of diseases, protection can be assured by maintaining high levels of hygiene, including keeping rodents out of dwellings. Fortunately, transmission is automatically limited because the ecological habits of humans and many infective rodent species have little or no overlap.

Jane Burton/Bruce Coleman Ltd

in elaborate tunnel systems.

Microtines have complex social structures. Scent is used to define male and female territories in the breeding season. According to the species, male and female territories may be separate or overlap; females may have separate or shared territories. Males, which can be monogamous or promiscuous according to the species, commonly establish dominance hierarchies. Dispersal is an important facet of microtine behavior. In some species, for example the meadow vole *Microtus pennsylvanicus*, the dispersers have a different genetic makeup, more suitable for establishment in

new locations, from the stay-at-homes. An expression of dispersal is the mass migrations of lemmings. In the Norway lemming *Lemmus lemmus*, these usually occur in summer or fall after a population buildup. Migrations start as small modest movements from upland areas. Initially, movement is random. As the descent progresses, more and more animals become incorporated into the migration until finally, with many thousands of animals involved, mass panic sets in. Movement becomes more reckless and undirectional, culminating in mass deaths from drowning in lakes, rivers, and even the sea.

▲ *The European water vole (usually known as the "water rat" in England) is primarily herbivorous, like other voles, but also eats insects, mice, and small birds. Related to rats and mice, voles have smaller eyes, a blunter snout, and a shorter tail.*

Jen & Des Bartlett/Bruce Coleman Ltd

▲ *The crested rat, which is the only member of the subfamily Lophiomyinae, inhabits arid, rocky areas of northeastern Africa. In response to threat, it has the very unusual ability to "part" tracts of hair along its body to reveal bold lateral stripes and scent glands.*

More African rats and mice

While most African rats and mice are in the subfamily Murinae, a further 51 species, excluding gerbils, are not included in this group. They are a diverse assemblage, ranging in size from the crested rat *Lophiomys imhausii* at 2.5 kilograms (5½ pounds) to Delany's swamp-mouse *Delanymys brooksi* at 5 to 7 grams (³⁄₁₆ to ¼ ounce). They are widespread throughout the continent south of the Sahara. One group of eleven species, the Nesomyinae, is restricted to Madagascar. Probably derived from a single ancestor, there are now species resembling mice, rats, voles, and rabbits.

The crested rat has long hair, a bushy tail, and the ability to part its hairs along the length of its body to expose its lateral stripes and scent glands. The other large species are the giant rats, *Cricetomys* species, which can weigh 1.2 kilograms (2½ pounds); they have cheek pouches, dull brown or gray dorsal fur, and a bicolored tail. Smaller species include the climbing mice, *Dendromus* species. Their small size—up to 18 grams (⅔ ounce)—and prehensile tails enable them to climb grasses and herbaceous vegetation. The swamp-rats (*Otomys*) inhabit grasslands, particularly in wet areas, where they can be extremely numerous. They are vole-like in appearance, are grass eaters, and, like voles, leave small piles of grass clippings in their runs.

Subterranean rats

There are 22 species of vegetarian, subterranean murids in Eastern Europe, Africa, and Asia. They inhabit burrows constructed with the aid of their protruding front teeth and strong forelimbs. Their food generally consists of roots, rhizomes, and bulbs, and underground storage is common. "Mole hills" are often constructed. The most highly adapted are the eight species of blind mole-rats, the Spalacinae, which have no external ears, no tail, and eyes permanently beneath the skin. The zokors, of the subfamily Myospalacinae, are less extremely adapted and have external ears, tails, and very small eyes. The root rats, of the subfamily Rhizomyinae, include the east African mole rats and the Asian bamboo rats, the latter showing the least underground specialization. Several species are serious agricultural pests.

Desert dwellers

Within this grouping are the hamsters, of the subfamily Cricetinae, the gerbils, of the family Gerbillinae, and the jerboas. The last are in a separate family, the Dipodidae, which has 11 genera and 30 species. With the exception of the hamsters, which also have a limited range in Europe, they are confined to Asia and Africa. They are found from temperate rocky mountains, steppes, and deserts to dry tropical grasslands. The jerboas differ from the hamsters and gerbils in having greatly extended hindlegs, and the main bones of the foot fused to form a single cannon bone. This adaptation enables jerboas to rapidly traverse remarkably long distances: a single animal has been recorded as covering over 12 kilometers (7½ miles) in one night.

Living in places with scant vegetation poses problems of predator avoidance. These problems are partially solved by exceptionally acute hearing through a highly developed middle ear; activity confined to the night; a burrow into which to bolt; excellent vision, often covering a broad field; and a body color that blends with the background. The desert dwellers are primarily granivores and vegetarians, though they often eat insects and other animals too.

In hot arid climates water intake and retention are particular problems. Gerbils and jerboas minimize water loss by sleeping in deep cool burrows during the day; in jerboas the depth of burrow varies seasonally according to the external temperature, the lower levels being favored in hot weather. Plugging the entrance prevents evaporation and maintains a moist internal environment. Water loss is also minimized by highly adapted kidneys that produce concentrated urine and digestive systems that produce dry feces.

▼ Most gerbils live in deserts or very arid country. They sleep in a deep, blocked-off burrow during the day and feed at night, mostly on seeds. They do not need to drink, obtaining sufficient water from their food and conserving this by excreting only small amounts of highly concentrated urine.

NHPA/Australasian Nature Transparencies

Dormice

Dormice occur in Africa south of the Sahara, from western Europe and North Africa to central temperate Asia and Arabia, and in Japan. With their agility, bushy tails, and typically arboreal habits, they share many characteristics with squirrels. The temperate species hibernate from about October to April. Mating takes place soon after emergence, with the first litters appearing in May. As they have no cecum, cellulose presumably plays little part in their diet. Mainly vegetarian, particularly the edible dormouse *Glis glis* and the common dormouse *Muscardinus avellanarius,* they eat fruits, nuts, seeds, and buds; the garden dormouse, *Eliomys* species, tree dormouse *Dryomys nitedula* and African dormouse, *Graphiurus* species, eat more animal material. Dormice can be very vocal: the sounds they make have been variously described as twitters, shrieks, clicks, growls, and whistles. Such behavior probably has territorial, mating, and hierarchical functions. *Selevinia betpakdalensis* is a desert dormouse of eastern Kazakhistan in the Soviet Union. Two species of spiny, oriental dormice are included in the murids.

Jumping mice and birch mice

Jumping mice are widespread through North America and Europe, and birch mice are found in Central Asia. They have poorly developed cheek pouches, very long tails and, in the jumping mice, particularly elongated hindfeet. Hibernation is prolonged, lasting from six to nine months. Body temperature may then fall to a little above freezing and bodily functions operate at minimal levels. The birch mice of the *Sicista* species, with their partly prehensile tails, are good climbers and spend much of their lives in bush and thicket. Jumping mice are ground dwelling.

◀ *The European dormouse.*

UPS & DOWNS OF RODENT POPULATIONS

J. Frazier/Australasian Nature Transparencies

▲ *The introduced house mouse erupts in plagues periodically in countries like Australia.*

Rodent populations are subject to considerable fluctuations in numbers. Among the best-known examples are the cyclical peaks of north temperate voles and lemmings, which occur every three to four years. In the increase phase of the cycle, the breeding season is protracted and breeding activity at a maximum. At the peak, the breeding season is abbreviated and juvenile mortality high. Food becomes short; stress sets in; and the population soon declines. A gradual recuperation of the population follows.

The underlying causes of these fluctuations are not firmly established. Two possibilities have been proposed. One is that growth and decline are attributable to exploitation and exhaustion of the food supply. The other is that behavioral changes within the population through the cycle maximize the species' ability to exploit its environment as well as colonize new areas.

In the semi-arid Sahel region of west Africa, the granivorous gerbil *Taterillus gracilis* underwent population fluctuations from less than 1 per hectare in 1971–73 to 143 per hectare in 1976. In this precarious environment, which has a low annual rainfall, climatic variations greatly influence plant productivity and rodent resources. Below-average rainfall in 1971 and a drought in 1972 reduced numbers. But above-average rainfall in 1972–76 meant more food and a longer breeding season. Another Sahelian species, the Nile rat, is normally present at low densities. It reached the incredible figure of 613 per hectare in 1975.

CAVY-LIKE RODENTS

The Caviomorpha derives its name from the South American guinea pig or cavy, which has a number of features in common with other members of the group, particularly in South and Central America. The characteristic jaw muscles, large size, robust body, large head, slender limbs, and short tail are widespread attributes, though only the first is ubiquitous.

The cavy-like rodents differ from squirrel-like and mouse-like rodents in having the largest number of families and the smallest number of species. In fact, four families are represented by only one species each. The reason for the more striking and fundamental diversification of the cavy-like rodents probably lies in their early evolution. It has been suggested that the first caviomorphs evolved from a Northern Hemisphere ancestor during the Eocene (57 to 37 million years ago), and that entry into South America took place across a sea barrier during the late Eocene or early Oligocene (37 to 24 million years ago). After North and South America reconnected in the Pliocene (5 to 2 million years ago) only one caviomorph, a porcupine, moved north and successfully established itself.

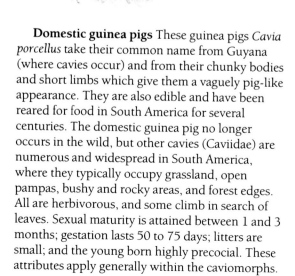

New World cavy-like rodents

Porcupines American porcupines (family Erethizontidae) are found from northern Argentina to Alaska and Canada. They have much shorter protective spines (about 4 centimeters, 1½ inches) than their Old World counterparts. Herbivorous, unaggressive, and slow moving, they spend much of their time in trees. Some, such as the prehensile-tailed and tree porcupines (*Coendou* species) are more arboreal than others. Normally, the spines lie smoothly along the back but at times of danger are erected by muscles in the skin. Quills are easily detached and remain embedded in an attacker by means of small barbs along much of their length. In the thin-spined porcupine *Chaetomys subspinosus* of the Brazilian forests the sharp quills are restricted to the head.

Domestic guinea pigs These guinea pigs *Cavia porcellus* take their common name from Guyana (where cavies occur) and from their chunky bodies and short limbs which give them a vaguely pig-like appearance. They are also edible and have been reared for food in South America for several centuries. The domestic guinea pig no longer occurs in the wild, but other cavies (Caviidae) are numerous and widespread in South America, where they typically occupy grassland, open pampas, bushy and rocky areas, and forest edges. All are herbivorous, and some climb in search of leaves. Sexual maturity is attained between 1 and 3 months; gestation lasts 50 to 75 days; litters are small; and the young born highly precocial. These attributes apply generally within the caviomorphs.

▲ *Prehensile-tailed porcupines (top) are arboreal leaf-eaters in tropical American forests. New World and Old World porcupines belong to different families and it seems that spines evolved independently in each group. The South American family Echimyidae, generally known as spiny rats (bottom), includes some soft-furred species but most have hair that is stiff and spiny. Although not at all closely related to porcupines, the spiny rats suggest how quills may have evolved from ordinary hairs.*

▲ Long-legged South American rodents. Top, the Patagonian mara or hare, a monogamous species that rears its young in a communal crèche. Center, the plains viscacha. Males of this species are twice the weight of females. Bottom, the golden agouti, a social species that shares a common burrow system.

food is abundant they may congregate in groups of up to 100, the pair-bonds being maintained.

Capybara The capybara *Hydrochoerus hydrochaeris* is the only living member of an apparently fairly recent family (Hydrochoeridae) whose fossil record only goes back to the Pliocene (5 million years ago). It is the largest living rodent and can weigh up to 66 kilograms (145 pounds), although the fossil members of the family were all considerably bigger than this extant species. Semi-aquatic, its adaptations include nostrils, eyes, and ears all high on the head. They graze at water's edge and are social animals living in groups of up to about 15, which may temporarily aggregate into larger assemblages. Capybaras are widespread east of the Andes from Panama to Argentina. In places they are cropped for food and leather. Another large herbivorous rodent, the coypu *Myocaster coypus* (family Myocastaridae), which weighs up to 10 kilograms (22 pounds), is more aquatic in its habits than the capybara. It burrows into river banks. Introduced into farms in various parts of the world for their fur, escaped coypus have established damaging feral populations.

Chinchillas and viscachas These members of the family Chinchillidae are medium-sized grazers that have dense coats and well-furred tails. The forelimbs are short and the hindlimbs long. With their large ears the mountain viscachas, of the genus *Lagidium*, look like rabbits. The plains viscacha *Lagostomus maximus* has short ears and is sexually dimorphic, the males being twice the size of the females. The smaller chinchillas, which weigh up to 800 grams (1¾ pounds), are prized for their fur.

Agoutis The agoutis (Dasyproctidae), which weigh up to about 2 kilograms (4½ pounds), have a short tail and long slim limbs adapted for rapid running. The forefeet have four toes and the hind three toes; the claws are blunt and almost hoof-like in some species. Diurnal forest-dwellers, they live in excavated burrows and feed on fallen fruits.

Christian Zuber/Bruce Coleman Ltd

▲ Capybaras, the largest living rodents.

Cavies These medium-sized rodents weighing 300 to 1,000 grams (10 to 35 ounces) are nocturnal and often live in colonies. Exceptions are the maras or Patagonian hares *Dolichotis patagonum* and *D. salinicolum*, which are much larger (8 to 9 kilograms, 17 to 20 pounds). With their long limbs they are capable of rapid running. In *D. patagonum*, male and female are monogamous for life. The male defends the female but does not assist in rearing the pups. Up to 15 pairs can deposit their young in a communal den where they remain for up to 4 months. Adults do not enter the den, but the pups leave it periodically for suckling and grazing. Communal suckling does not take place. Generally adults are dispersed in pairs but when

Hutias The 13 species of hutias (Capromyidae) are restricted to the West Indies, where they have evolved to fill a number of niches. They are heavy, thickset rodents, the largest being the forest-dwelling hutia *Capromys pilorides,* which weighs up to 7 kilograms (15 pounds). Although arboreal it has, like other hutias, an underground den.

Tuco-tucos, degus, and coruros The tuco-tucos (Ctenomyidae) of the lowland pampas and upland plateaus are the South American equivalent of the pocket gophers. Both are adapted for digging. The fur is dark to pale brown; the limbs are short with large digging claws and fringes of hair on the toes and limbs to assist in moving soil; the tail is short; and the eyes and ears small. Leaves, stems, and roots are eaten. Burrowing is also common in the degus (Octodontidae), which occur from sea level to 5,000 meters (16,000 feet). One of the more specialized species is the coruros *Spalacopus cyanus* of Chile. Groups of animals occupy a common burrow system where they feed entirely underground, largely on the tubers and stems of a particular lily. When the food supply is exhausted the colony moves to a new site.

Spiny rats The spiny rats (Echimyidae) are an assemblage of 55 species of medium-sized herbivores covered with spiny or semi-spiny bristles. Superficially similar to the murids, they were established in South America several million years before this group entered. Within their range from Nicaragua to Paraguay they can be very common in both forest and grassland. The species with long tails and slender bodies tend to be tree dwellers; the more solid short-tailed species are ground-dwelling burrowers; and the intermediate forms exploit both situations. These animals readily lose their tail owing to structural weakness in the fifth caudal vertebra—doubtless a unique adaptation to escape a predator.

Old World cavy-like rodents

As in their South American counterparts, there is a considerable range of form and size within this group. The smallest is the naked mole-rat of East Africa *Heterocephalus glaber,* which weighs 30 to 40 grams (1 to 1⅓ ounces), and the largest, the crested porcupine *Hystrix,* which weighs up to 24 kilograms (53 pounds). The large African cane rats *Thryonomys* species, which weigh up to 9 kilograms (20 pounds), have a covering of short, harsh, bristly hairs. They prefer thick grassland close to water and the larger species, *T. swinderianus,* is semi-aquatic. They feed mainly on grasses but are renowned as pests of maize, sugar cane, and other agricultural crops. On the other hand, they are very popular as a source of food.

Porcupines Included in the Old World porcupines (Hystricidae) are three species with long tails and a distinct terminal tuft of bristles. In the African (*Atherurus africanus*) and Asian (*A. macrourus*) brush-tailed porcupines, each bristle

consists of a chain of flattened discs, enabling the tail to be rattled, probably as a warning to predators. These relatively small (up to 3.5 kilograms, 7¾ pounds), forest-dwelling species lie up by day in burrows, crevices, and caves. Roots, tubers, and fallen fruit are their main foods. The long-tailed porcupine *Trichys lipura* of Malaya, Sumatra, and Borneo has a terminal brush of hollow bristles. It is small (1.5 to 2 kilograms, 3 to 4½ pounds) and, with a body covered in short bristles, looks like a spiny rat.

The remaining porcupines, *Hystrix* species, have shorter tails and longer quills. They occur from Java and Borneo, through southern Asia to southern Europe, and through most of Africa. They are vegetarian, favoring roots, bulbs, and fruit. They excavate burrows, which they share with their progeny. If they are attacked, the quills of the back are raised and those on the tail rattled as a threat. The porcupine then attacks by running sideways or backwards, trying to drive its spines into its adversary.

Gundis Of the family Ctenodactylidae, gundis are small rodents, weighing up to 180 grams (about 6 ounces), that live in broken, rocky country in northern and northeast Africa. They are agile and dexterous rock climbers, aided by the sharp claws and pads on the soles of their feet. Their soft fur insulates the body from excessive exposure to the sun—an important adaptation for species foraging during the warm times of the day when the temperature ranges from 20° to 30°C (68° to 90°F). At the hottest part of the day they shelter from the sun on shaded rock ledges. Their sharp nails are unsuitable for grooming the soft coat; rows of stiff bristles arising from the two inner toes of the hindfeet are used instead. Gundis live in groups within which are family territories. They are opportunistic herbivores with a limited capacity to concentrate their urine when dry food is eaten.

▲ *The crested porcupine. Like other Old World porcupines, it is terrestrial and has a relatively small and non-prehensile tail. The long spines (quills) are erected and rattled in response to threat, and the porcupine then rushes backward or to the side, attempting to drive them into its attacker.*

John Visser/Bruce Coleman Ltd

▲ *Like most mammals that live underground, mole-rats are blind. The legs are short and powerful and the body is cylindrical.*

Mole-rats As their common name suggests, mole-rats (Bathyergidae) are rats that have assumed habits similar to moles. The nine species, all occurring in Africa, display considerable adaptation to subterranean life, often paralleling their murid counterparts. They differ from the insectivorous moles in being vegetarian. Mole-rats spend almost their entire lives underground, inhabiting burrow networks of varying complexity. The lips close behind the incisors to prevent soil entering the mouth. The soil is then pushed back by the hindfeet, and when a pile accumulates it is removed by the rat to the outside to appear as a "mole hill" or to be displaced elsewhere in the burrow network. Further adaptations for a subterranean life include cylindrical bodies, short tails, small eyes which no longer serve the function of sight but have instead become sensitive to air currents in the burrow, loss of outer ears, and keen senses of hearing and smell. Mole-rats range from 30 to 60 grams (1 to 2 ounces) in the hairless naked mole-rat *Heterocepahlus glaber*, to about 800 grams (28 ounces) in the Cape dune mole-rat *Bathyergus suillus*. Mole-rats feed on roots, tubers, and the like obtained as they excavate. Their burrowing has been known to cause railway lines to sag.

M. J. DELANY

THE NAKED MOLE-RAT: A HIGHLY SOCIAL RODENT

Jane Burton/Bruce Coleman Ltd

▲ *Except for sensitive bristles on the snout, the naked mole-rat has no hair. Its naked condition may have evolved as a means of losing heat from the body.*

The fascinating social organization of the naked mole-rat is unique among mammals. These almost hairless rodents live in colonies of up to 40 (and possibly more) individuals in the drier, warmer savannas of East Africa. They inhabit an elaborate network of foraging tunnels plus a breeding chamber. In the naked mole-rat colony there are clear divisions of labor, presumably to make the community function more efficiently, which have parallels only in the social insects such as bees, wasps, and ants.

Within the colony a single breeding female produces recognizable castes of the more numerous "frequent workers" and the less numerous non-workers. The former are the smallest members of the colony; they dig, transport soil, forage, and carry food and bedding to the communal nest. Digging is performed cooperatively: each animal excavates its soil, then pushes it backwards under the forward-moving animals who have already disposed of their load. Underground food supplies are exposed during the course of this digging.

The larger non-workers spend most of their time in the nest with the breeding female. The females are not sexually active, but they have the potential to be so if the breeding female is removed. The breeding female produces 1 to 4 litters a year with an average of 12 young, though up to 27 have been recorded. The young are suckled only by the breeding female, and at weaning eat food brought to the nest and feces of colony members. After weaning growth is slow. The juveniles join the worker caste, but some eventually grow larger to join the non-worker caste. It is probable that a hierarchy exists within the non-workers, dominated and controlled by the breeding female. Chemicals (pheromones) secreted by her probably inhibit breeding in non-workers. Chemical influence by the breeding female is also witnessed immediately before she gives birth, when all colony members develop teats and male hormone levels drop. This is probably to prepare the colony to care for the young.

LAGOMORPHS

KEY FACTS

ORDER LAGOMORPHA
- 2 families • 13 genera
- *c.* 80 species

SMALLEST & LARGEST

Steppe pika *Ochotona pusilla*
Head–body length: 18 cm (7 in)
Weight: 75–210 g (2½–7½ oz)

European hare *Lepus europaeus*
Head–body length: 50–76 cm
(20–30 in)
Weight: 2.5–5 kg (5½–11 lb)

CONSERVATION WATCH

!!! The Helan Shan pika *Ochotona helanshanensis*; Sumatran rabbit *Nesolagus netscheri*; and Omiltemi rabbit *Sylvilagus insonus* are critically endangered.
!! 8 species are endangered, including: Koslov's pika *Ochotona koslowi*; riverine rabbit *Bunolagus monticularis*; Dice's cottontail *Sylvilagus dicei*.
! 5 species are listed as vulnerable.

The order Lagomorpha includes hares, rabbits, and pikas — a familiar group of animals with large ears and eyes set wide on their heads. Lagomorphs have a widespread distribution, either naturally or as a result of introduction by humans, but they are absent from parts of Southeast Asia, including several of the larger islands. Although typically animals of tundras, open grasslands, rocky terrains, and arid steppes, some occur in temperate woodlands, cold northern forests, and tropical forests.

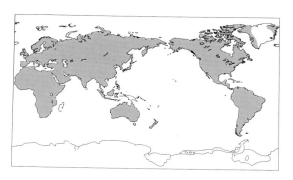

PIKAS

Pikas, which belong to the family Ochotonidae, are the smallest lagomorphs, weighing from 80 to 300 grams (3 to 11 ounces). They have prominent rounded ears and no tail. They occur in steppes and rocky areas, particularly slopes, up to 6,000 meters (20,000 feet) in western North America (two species) and northeastern Asia (twelve species). Pikas retire to rock crevices or burrows according to the terrain. Some species, for example the Afghan pika *Ochotona rufescens* and the daurian pika *O. daurica*, are found in both types of terrain; others, such as the steppe pika *O. pusilla*, are burrowing steppe-dwellers. Vertical segregation of species is common, particularly on high mountains; even on low ground there is little evidence of two species occupying the same habitat.

Pikas are opportunistic vegetarians. They eat any plants available near their homes. During summer and autumn they harvest plants and hoard them under rocks or in burrows for the winter. These hay piles are stored within a territory, which in the case of the northern pika *O. hyperborea* is held by a male and a female throughout the year.

HARES AND RABBITS

Hares and rabbits belong to the family Leporidae. They range in size from the pygmy rabbit *Sylvilagus idahoensis*, which weighs up to 300 grams (11 ounces), to the Arctic hare *Lepus timidus*, which weighs up to 5 kilograms (11 pounds).

In several parts of the world, members of this family are best known for the damage and depredations wrought by the European rabbit *Oryctolagus cuniculus*, which was introduced by the Normans into Britain, and from there by eighteenth and nineteenth century settlers into Australia and New Zealand. Once the European rabbit was established in the wild in these new environments, its numbers increased and livestock pastures and crops were devastated. In Australia, the near extinction of certain marsupials can be attributed to competition with rabbits.

Rabbits, including the European rabbit, can be prolific breeders. One doe can produce up to thirty young in a single breeding season, from January to August. Even though rabbits are born blind, inactive, and with a thin covering of fur, they are capable of breeding at three and a half months.

Some species, however, exist in relatively small numbers. The volcano rabbit *Romerolagus diazi* is restricted to two volcanic sierras close to Mexico City; the Amami rabbit *Pentolagus furnessi* to two of the Japanese Amami islands; and the little-known greater red rockhare *Pronolagus crassicaudatus* is found only in parts of Natal and Cape Province, South Africa.

▲ Although they resemble rodents such as guinea pigs or hamsters, pikas are most closely related to rabbits and hares. Like these, they have two pairs of upper incisors and a slit between the upper lip and the nostrils ("hare lip"). Pikas store leaves and grass during the summer to be eaten, as hay, in winter.

Martin W. Grosnick/Ardea London

John Cancalosi/Auscape International

► In addition to their function in hearing, the enormous ears of the black-tailed jack rabbit are radiators that help to control the temperature of this desert animal. When the blood vessels of the ears are engorged with blood, the jack rabbit loses heat; when they are constricted, body heat is conserved.

▼ The "boxing" that occurs between European hares seems to be mainly a matter of a female rejecting an over-amorous male.

it. During breeding, dominant does have their nesting chambers in the main warren while the lowest-ranking females have to nest in less-favored locations. Territories are marked by secretions from a gland under the chin and the anal glands, and by urine. These are important components of communication by smell.

In Europe, foxes and stoats are among the main predators of adult rabbits; the young are vulnerable to a wider range of predators including birds of prey, weasels, badgers, and domestic cats. Alarm signals are provided by the rabbit's flashing white tail when it runs, and by thumping with the feet both above and below ground. Both rabbits and hares indulge in tooth-grinding, which may serve as a warning signal. With their long, strong hindlegs, hares and rabbits are adapted for rapid running. Agility, plus their protective coloring, keen sense of smell, and long mobile ears which give them acute hearing, are their main defenses against predators. In contrast to the highly social rabbit, many hares, including the European brown hare, are solitary. They are seen in pairs or small groups only during the breeding season.

Hares and jack rabbits of the *Lepus* genus differ from the remaining nine genera (some of which are confusingly called hares) in several respects. They do not burrow much and have developed improved running ability. Hares typically give birth to small litters of well-furred, mobile young,

The European rabbit can be active at any time of the day or night. It lives in groups with a highly organized social structure maintained, particularly at high densities, by aggressive conflict. Dominant bucks and does occupy the whole territory, while subordinates are restricted to smaller areas within

Stefan Meyers/Ardea London

Hans Reinhard/Bruce Coleman Ltd

Hans Reinhard/Bruce Coleman Ltd

which have open eyes. Litters are deposited on a form, that is, the hare's bed, shaped by the animal's body. The European brown hare *Lepus capensis* can breed throughout the year, although the main reproductive activity is in April and May. A female can produce three or four litters a year, totalling about seven to ten young — a much lower rate of production than that of the rabbit.

The spectacular "boxing matches" and chasing seen in brown hares in spring has given rise to the phrase "mad as a March hare". This activity can hardly be related to breeding, which starts appreciably earlier than March and continues well into winter. Much of the "boxing" is probably the doe repelling an over-ardent buck. And, as chasing, fighting, and mating go on for much of the year, it appears that the concept of "March madness" has arisen from limited observation!

The Arctic hare is well adapted to living in the far north, as its white winter coat provides excellent concealment. In summer, the coat changes to brown or grey. It is one of a few species of hare that will dig a burrow—into snow or into

earth. It can be used as a bolt-hole for the young, as the adults rarely enter. Another species of the north with a white winter coat is the snowshoe hare *Lepus americanus*. While the Arctic hare extends across much of the Northern Hemisphere, the snowshoe hare is restricted to North America. The value of this animal's pelt resulted in its systematic trapping for over two centuries.

Two species of hare are widespread in Africa, occurring everywhere but in tropical rainforest areas and the most arid parts of the Sahara. They are the Cape hare *Lepus capensis* and the scrub hare *Lepus saxatilis*. Both are nocturnal, lying up in their forms during daytime, and both are grass eaters, but there is a sharp distinction in their habitat preferences. The Cape hare is found in more arid, open terrain, and the scrub hare in scrub and woodland adjacent to grassland. That the former is well adapted to arid country is confirmed by its success in Arabia. Here it is found in sand desert, steppe, scrubland, and mountains, wherever there is an adequate food supply.

M. J. DELANY

▲ The Arctic or alpine hare has a brown or gray coat in summer (above, left), which changes to white in winter (above, right). The snowshoe hare undergoes the same seasonal change, which aids concealment. However, a similar change in the coats of the Arctic fox and snowy owl also reduces the visibility of these predators.

ELEPHANT SHREWS

SMALLEST & LARGEST

Short-eared elephant shrew
Macroscelides proboscideus
Head–tail length: 23.5 cm (9 in)
Weight: 45 g (1½ oz)

Golden-rumped elephant shrew
Rhynchocyon chrysopygus
Head–tail length: 56 cm (22 in)
Weight: 540 g (19 oz)

CONSERVATION WATCH
!! 3 species are endangered,
including the golden-rumped
elephant shrew *Rhynchocyon
chrysopygus*.
! 4 species are vulnerable.

▼ *Elephant shrews use their long,
sensitive, and mobile snout to probe leaf
litter for invertebrate prey. This spectacled
elephant shrew is eating a cricket.*

Elephant shrews are a highly distinctive group of animals. They are so named because of the resemblance—vague though it is—between the trunk of an elephant and the extended, highly flexible snouts of these animals. All the elephant shrews are characterized by long, thin legs, large eyes, external ears, long, rat-like tails, and, of course, the proboscis-like snout.

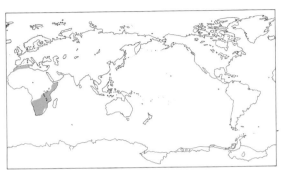

SPECIALIZED BOUNDERS

In comparison with the insectivores, elephant shrews have a well-developed brain. In body form they resemble the oriental tree shrews more closely than any other living group, but they have specialized in a totally different way of life. Almost all of their time is spent on the ground, where their long legs enable them to jump and run at impressive speeds. Alone among insect-eating mammals they are specialized for bounding, which is probably a useful anti-predatory tactic that serves to confuse pursuing predators.

Elephant shrews occupy a considerable variety of habitats, including open plains, savannas, thornbush, and tropical forests. Although many of the species are active during at least part of the day, and are therefore fairly easy to see, their behavior and ecology remain poorly known.

TYPES OF ELEPHANT SHREW

As with the true insectivores, the classification of the elephant shrews has had a checkered history. At one time or another, elephant shrews have been associated with primates, insectivores, tree shrews, and even ungulates. Now, however, elephant shrews have been firmly installed in their own distinct order, Macroscelidea, which has one family, the Macroscelididae. Within this family there are four genera: the *Macroscelides* (one species), *Petrodromus* (one species), *Elephantulus* (ten species), and *Rhynchocyon* (three species).

Body form varies considerably according to species. The aptly named golden-rumped elephant shrew *Rhychocyon chrysopygus*, for example, is a large species, measuring 56 centimeters (22 inches) from the nose to the tip of the tail and weighing approximately 540 grams (19 ounces). These animals are a striking dark amber color with black legs and feet and a distinct gold-colored rump patch. The short-eared elephant shrew *Macroscelides proboscideus*, in contrast, is a dull gray-brown color and is the smallest species of the group, measuring only 23.5 centimeters (9 inches) and weighing about 45 grams (1½ ounces). There is no sexual dimorphism within a given species.

TERRITORY AND NESTS

Elephant shrews have a clearly defined monogamous mating system. Animals display a strong degree of fidelity toward their partners—they usually mate for life—and, together, each pair defends a fixed territory against other elephant shrews. When faced with intruders, the resident male usually engages in contest if the intruder is a male, the female of the pair if the intruder is a female. Direct interactions between the male and female of a pair are rare, but contact is

Jane Burton/Bruce Coleman Limited

achieved through depositing scent at strategic parts of the shared territory.

At least one species, the spectacled elephant shrew *Elephantulus rufescens,* builds and maintains an elaborate system of trails traversing the territory. The male of the pair may spend as much as 25 percent of his active time clearing debris such as leaves off the trail system by pushing it away with his forepaws. Grasses and small branches are chewed until they break into pieces small enough to move. The reason for such fastidious housework appears to be that it provides clear escape routes within the home territory if any family member is surprised by predators. The trail system may often extend for more than several hundred meters and remains relatively stable throughout the year. All activities, such as resting, sleeping, grooming, or foraging, occur at favorite parts of the trail system.

The spectacled tree shrew does not construct a complete nest, but some other species create elaborate nests and burrows. Nest sites are dispersed throughout the range, usually at the base of a tree or under a heap of branches. When they sleep, elephant shrews rarely lie on their sides; instead, they keep their limbs beneath the body in a position that allows for a quick escape.

FINDING FOOD

Apart from serving as a communications highway, the trail system is used extensively for foraging. The principal food items are ants and termites, although beetles, worms, leaf litter, invertebrates, and, occasionally, small lizards are captured.

The various species have different techniques for locating prey: species of the genus *Rhynchocyon,* for example, appear to detect potential prey by smell as they forage among leaf litter, continuously poking their long snouts into the debris and thereby seeking out beetles, grasshoppers, and spiders. In contrast, the spectacled elephant shrew prefers to feed on termites and ants, breaking into their tunnel systems by a combination of clawing and biting. Because of the dispersed, unpredictable nature of leaf-litter invertebrates, there is little advantage for an animal feeding on them to defend such a resource. Thus, the golden-rumped elephant shrew spends about three-quarters of its active time looking for food within a territory of almost 2 hectares (5 acres), but the spectacled elephant shrew, which feeds primarily on termites, a localized food resource, spends less time searching for prey and so needs to defend a smaller territory, usually less than half a hectare (1 acre).

BREEDING BEHAVIOR

Elephant shrews breed throughout the year. At a carefully chosen site, females give birth to one or two rather precocial young which are nursed only at 24-hour intervals. Because the infants are so well developed they require minimal parental attention in terms of feeding, warmth, protection, and

grooming, and thus do not serve as a focus for maternal and paternal behavior. This behavior is probably intended to reduce any attention from potential predators. Juveniles remain at the nest site for about two weeks, after which they begin to forage independently, with only infrequent nursing. Unlike insectivores such as moles, shrews, or desmans, weaned elephant shrews are permitted to remain within the adults' territory until they reach sexual maturity, when they must seek a vacant territory.

The most vulnerable stage of the life cycle is probably that experienced by the newly independent subadult animals, as they do not have an established territory. The main predators of elephant shrews are snakes, owls, and small carnivores. In some areas of East Africa, species of the genus *Rhynchocyon* are eaten by local people and are hunted with dogs for this purpose. Few animals live to more than four years of age, when tooth wear becomes a serious handicap to feeding.

CONSERVATION

Nearly every species is associated with a specific, often narrow-ranged, habitat. For example, the rock elephant shrew *Elephantus myurus* lives almost exclusively on rocky outcrops, whereas the largest species, the golden-rumped elephant shrew, frequents the periphery of dry deciduous forest. In such environments, which are subject to human interference, these and other species may not be adaptable enough to survive the destruction of their habitats. More than half of all known species are now recognized as either endangered or vulnerable.

R. DAVID STONE

Jen & Des Bartlett/Bruce Coleman Limited

▲ *Strictly terrestrial, elephant shrews have long, slender legs that enable them to avoid predators by running fast or bounding like miniature antelopes. Species range in size from that of a mouse to that of a half-grown rabbit.*

FURTHER READING

● **INTRODUCING MAMMALS (P. 14)**

Anderson, S., & J.K. Jones, Jr, 1984. *Orders and Families of Recent Mammals of the World.* Wiley, New York.

Macdonald, D., 1984. *The Encyclopedia of Mammals.* Facts on File Inc., Allen & Unwin, London.

Nowak, R.M., & J.L. Paradiso, 1983. *Walker's Mammals of the World,* 4th edn. Johns Hopkins, Baltimore.

Vaughan, T.A., 1972. *Mammalogy.* W.B. Saunders Co., Philadelphia.

Young, J.Z., 1975. *The Life of Mammals.* Oxford.

● **CLASSIFYING MAMMALS (P. 19)**

Corbet, G.B., & J.E. Hill, 1991. *World List of Mammalian Species,* 3rd edn. Natural History Museum, London.

Lawlor, T.E., 1979. *Handbook to the Orders and Families of Living Mammals.* Mad River Press, Eureka, California.

Wilson, D.E., & D.M. Reeder, 1993. *Mammal Species of the World,* 2nd edn. Smithsonian Institution, Washington.

● **MAMMALS THROUGH THE AGES (P. 22)**

Savage, R.J.G, & M.R. Long, 1986. *Mammal Evolution.* British Museum, London.

● **HABITATS & ADAPTATIONS (P. 28)**

Byers, J.A., 1998. *American Pronghorn: Social Adaptations and the Ghosts of Predators.* University of Chicago Press, Chicago.

Degen, A., 1996. *Ecophysiology of Small Desert Mammals.* Springer Verlag, New York.

Eisenberg, J.F., 1981. *The Mammalian Radiations.* University of Chicago Press, Chicago.

Marchand, P.J., 1991. *Life in the Cold: an Introduction to Winter Ecology,* 2nd edn. University Press of New England, Hanover, NH.

● **MAMMAL BEHAVIOR (P. 33)**

Manning, A., & M.S. Dawkins, 1998. *An Introduction to Animal Behaviour,* 5th edn. Cambridge University Press, Cambridge.

Smith, J.P., 1990. *Mammalian Behaviour: the Theory and the Science.* Bench Mark Books, Tuckahoe, New York.

Wright, R., 1994. *The Moral Animal.* Abacus Books, New York.

● **ENDANGERED SPECIES (P. 38)**

Burton, J.A., & B. Pearson, 1987. *Collins Guide to Rare Mammals of the World.* Collins, London.

Fisher, J., N. Simon & J. Vincent, 1969. *The Red Book: Wildlife in Danger.* Collins, London.

● **MONOTREMES (P. 48)**

Augee, M., & B. Gooden, 1993. *Echidnas of Australia and New Guinea.* UNSW Press, Sydney.

Grant, T., 1995. *The Platypus: a Unique Mammal.* UNSW Press, Sydney.

● **MARSUPIALS (P. 52)**

Dawson,T.J., 1983. *Monotremes and Marsupials: the Other Mammals.* Arnold, London.

Dawson, T.J., 1995. *Kangaroos: Biology of the Largest Marsupials.* UNSW Press, Sydney.

Hume, I.D., 1982. *Digestive Physiology and Nutrition of Marsupials.* Cambridge University Press, Cambridge.

Lee, A.K., & A. Cockburn, 1985. *Evolutionary Ecology of Marsupials.* Cambridge University Press, Cambridge.

Tyndale-Biscoe, H., & M. Renfree, 1987. *Reproductive Physiology of Marsupials.* Cambridge University Press, Cambridge.

● **ANTEATERS, SLOTHS & ARMADILLOS (P. 66)**

Montgomery, G.G., 1985. *The Evolution and Ecology of Sloths, Anteaters and Armadillos.* Smithsonian Institution, Washington.

Simpson, G.G., 1980. *Splendid Isolation.* Yale University Press, New Haven.

● **INSECTIVORES (P. 74)**

Churchfield, S., 1990. *The Natural History of Shrews.* Cornell University Press, Ithaca.

Eisenberg, J.F., & E. Gould, 1970. *The Tenrecs: a Study in Mammalian Behavior and Evolution.* Smithsonian Institution, Washington.

Godfrey, G.K., & P. Crowcroft, 1960. *The Life of the Mole* (T. europaea Linnaeus). Museum Press, London.

Gorman, M.L., & R.D. Stone, 1990. *The Natural History of Moles.* Christopher Helm, London.

Reeve, N., 1994. *Hedgehogs.* T&AD Poyner, London.

● **TREE SHREWS (P. 86)**

Stone, R.D., 1994. *Eurasian Insectivores and Tree Shrews: Status Survey and Conservation.* IUCN, Cambridge.

● **FLYING LEMURS (P. 88)**

Medway, L., 1978. *The Wild Mammals of Malaya and Singapore,* 2nd edn. Oxford University Press, Kuala Lumpur.

● **BATS (P. 90)**

Findley, J.S., 1993. *Bats: a Community Perspective.* Cambridge University Press, Cambridge.

Hill, J.E., & J.D. Smith, 1984. *Bats: a Natural History.* Rigby, London.

Kunz, T.H., 1982. *Ecology of Bats.* Plenum, New York.

Wilson, D.E., 1997. *Bats in Question.* Smithsonian Institution Press, Washington.

● **PRIMATES (P. 108)**

Altmann, S.A., & J. Altmann, 1970. *Baboon Ecology.* University of Chicago Press, Chicago.

Clutton-Brock, T.H., (ed.), 1977. *Primate Sociology.* Academic Press, London.

Fleagle, J.G., 1988. *Primate Adaptation and Evolution.* Academic Press, London.

Rainier, HSH Prince of Monaco, & G.H. Bourne, (eds), 1977. *Primate Conservation.* Academic Press, London.

Rowe, N., 1997. *Pictorial Guide to Living Primates.* Pogonias Press, New York.

Smuts, F., D.L. Cheney, R.M. Setfarth, R.W. Wrangham & T.T. Struhsaker (eds), 1986. *Primate Societies.* University of Chicago Press, Chicago.

• CARNIVORES (P. 134)
Bekoff, M., 1978. *Coyotes: Biology, Behavior and Management.* Academic Press, New York.

Bertram, B., 1978. *Pride of Lions.* Charles Scribner, London.

Ewer, R.F., 1973. *The Carnivores.* Wiedenfeld & Nicholson, London.

Fox, M.W., (ed.), 1978. *The Wild Canids: Their Systematics, Behavioural Ecology and Evolution.* Van Nostrand Reinhard, London.

Guggisberg, C.A.W., 1975. *Wild Cats of the World.* David & Charles, Newton Abbot, England.

King, C., 1990. *The Natural History of Weasels and Stoats.* Comstock, Ithaca.

King, J.E., 1983. *Seals of the World.* British Museum (Natural History), London.

Kitchener, A., 1991. *The Natural History of the Wild Cats.* Comstock and Cornell University Press, Ithaca.

• WHALES & DOLPHINS (P. 156)
Baker, M.L., l987. *Whales, Dolphins and Porpoises of the World.* Doubleday, New York.

Carwardine, M., 1995. *Whales, Dolphins and Porpoises.* Dorling Kindersley, London.

Evans, P.G.H., 1987. *The Natural History of Whales and Dolphins.* Christopher Helm Ltd, Kent.

Harrison, R.J., & M.M. Bryden (eds), 1988. *Whales, Dolphins and Porpoises.* Golden Press, Sydney.

Matthews, L.H., 1975. *The Whale.* Allen & Unwin, London.

Slijper, E.J., 1979. *Whales.* Hutchinson & Co, London.

• SEA COWS (P. 164)
Marsh, H., 1981. *The Dugong.* James Cook University Press, Townsville.

• ELEPHANTS (P. 167)
Delort, R., 1990. *Les Eléphant Piliers du Monde.* Découvertes Gallimard, Histoire Naturalles, Evreux, France. [English translation published 1992 by Abrams, New York, as *The Life and Lore of the Elephant.*]

Eltringham, S.K., & D. Ward (eds), 1991. *The Illustrated Encyclopedia of Elephants.* Salamander Books, London.

Shoshani, J., 1992. *Elephants: Majestic Creatures of the Wild.* Simon & Schuster, London.

Sukumar, R., 1989. *The Asian Elephant: Ecology and Management.* Cambridge University Press, Cambridge. 251 pp.

• ODD-TOED UNGULATES (P. 174)
Berger, J., 1986. *Wild Horses of the Great Basin: Social Competition and Population Size.* University of Chicago Press, Chicago.

Duncan, P., 1991. *Horses and Grasses: a Study of Horses and their Impact on the Camargue.* Springer, New York.

Owen-Smith, R.N., 1988. *Megaherbivores: the Influence of Very Large Body Size on Ecology.* Cambridge University Press, Cambridge.

Rhinoceroses. Meeting the Giants. Filander Verlag GmbH, Furth, Germany, 1997.

• HYRAXES (P. 188)
Maloiy, G.M.O., & R.M. Eley, 1994. *The Hyrax.* Regal Press, Nairobi, Kenya.

Meltzer, A., & M. Livneh, 1982. *The Rock Hyrax.* Massada Ltd., Giv'ataim, Israel.

• AARDVARK (P. 190)
Melton, D.A., 1974. 'The aardvark at night', *Animals,* London, 16:108-110.

Shoshani, J., C.A. Goldman, & J.G.M. Thewissen, 1988. '*Orycteropus afer*', *Mammalian Species,* 300:1-8.

• EVEN-TOED UNGULATES (P. 192)
Dagg, A.I., & J.B. Foster, 1976. *The Giraffe: Its Biology, Behavior and Ecology.* Van Nostrand Reinhold, New York.

Gaulthier-Pilters, H., & A.I. Dagg, 1981. *The Camel: Its Evolution, Ecology, Behavior and Relationship to Man.* University of Chicago Press, Chicago.

Geist, V., 1971. *Mountain Sheep: a Study in Behavior and Evolution.* University of Chicago Press, Chicago.

Schaller, G., 1967. *The Deer and the Tiger.* University of Chicago Press, Chicago.

Whitehead, G.K., 1972. *Deer of the World.* Constable, London.

• PANGOLINS (P. 212)
Heath, M.E., 1992. '*Manis crassicaudata*', *Mammalian Species,* 513:1-6.

• RODENTS (P. 212)
Barnett, S.A., 1975. *The Rat: a Study in Behaviour.* University of Chicago Press, Chicago.

Berry, R.J., (ed.), 1981. *Biology of the House Mouse.* Academic Press, London.

Hoogland, J.L., 1995. *The Black-tailed Prairie Dog: Social Life of a Burrowing Mammal.* University of Chicago Press, Chicago.

Sherman, P.W., J.U.M. Jarvis & R.D. Alexander, 1991. *The Biology of the Naked Mole Rat.* Princeton University Press, Princeton, New Jersey.

• LAGOMORPHS (P. 229)
Lockley, R.M., 1976. *The Private Life of the Rabbit.* Andre Deutsch, London.

Orr, R.T., 1977. *The Little-known Pika.* Collier MacMillan, New York.

Sheail, J., 1971. *Rabbits and their History.* David & Charles, Newton Abbot, England.

• ELEPHANT SHREWS (P. 232)
Nicholl, M.E., & G.B. Rathbun, 1990. *African Insectivora and Elephant-Shrews: an Action Plan for their Conservation.* IUCN, Gland, Switzerland.

INDEX

Page numbers in *italics* refer to photographs and illustrations.

A

aardvarks 19, 190–1, *190–1*
aardwolves *146*, 147
Acinonyx jubatus 134
Acomys 219
  A. dimidiatus 214
Acrobates 63
  A. pymaeus 57
Acrobatidae 57, 61
addaxes 41, 193
Aepomys 219
Agouti paca 214
agoutis 214, 226, *226*
Ailuropoda melanoleuca 134
Ailurus fulgens 134
Akodon 219
Allenopithecus 117
Allocebus trichotis 108, 112
Alopex 136
  A. lagopus 19, 137
Alouatta 115
  A. palliata 115
  A. seniculus 115
Amazon basin 40–1
Amblysomus hottentotus 77
Ameridelphians 53
angwantibos 110, *111*
anoas 207
Anotomys 214
anteaters 20, 26, 28, 66–8, 66–8, 73
Antechinomys laniger 54
antechinus 37
antelopes 26, 41
  African 179
  bongos 207
  bonteboks *210*
  duikers *210*
  dwarf 206
  four-horned 206, *207*
  gemsboks *210*
  giant sable *210*
  Hunter's 192
  Indian *209*
  klipspringers *210*
  sable 206
  saigas 193, *209*
  twisted-horned 206
  Uganda kob 35, 36
  see also gazelles; goat-
      antelopes;
      wildebeests
Antilocapridae 192
Aonyx cinerea 143
Aotus 115, 133
  A. trivirgatus 115
aoudads *211*
apes 26, 27, 108, *116*, 120–31
  chimpanzees 36, 108, 120,
      128–9, 128–31
  gibbons 108, 120, 122–3,
      122–3, 129, 130,
      131
  gorillas 44, *45*, 108, 120,
      120, 126–7,
      126–7, 130–1,
      130–1

orang utans 33, 108, 120,
      124–5, *124–5*, 129,
      130–1
Aplodontia rufa 215
Apodemus 219
Aproteles bulmerae 90
Arctocephalus 150, 151, 152
  A. australis 152
  A. galapagoensis 134
  A. pusillus pusillus 152
  A. tropicalis 152
armadillos 26, 66, 70, 72–3,
      72–3
Arsinoitherium 25
Artibeus 96, 103
Artiodactyla 21, 192
artiodactyls 25, 26, 192–3,
      206
Arvicanthis niloticus 220
asses 174, 176, *177*, 178
Ateles 115
Atherurus
  A. africanus 227
  A. macrourus 227
Australian mammals 27
Australidelphia 53
Australidelphians 54, 56
Australopithecus 26
Avahi 110
aye-ayes 61, 108, 110, *111*,
      112, 120

B

babirusas *194*, 197
baboons 30, 34, 37, 108, 117,
      117
backbones 17
badgers 141, 142–3
baijis 156
Balaena mysticetus 156
Balaenidae 157
Balaenoptera
  B. borealis 156
  B. musculus 156
  B. physalus 156
Balaenopteridae 157
baleen 158
bandicoots 36, 55, 56, 61, 63,
      65
  Australian 33
  northern brown *55*
  pig-footed 55
  rabbit-eared 55
  spiny 55
  western barred 52
Barbary apes 4–5, 118
Bassaricyon gabbii 140
Bathyergidae 228
Bathyergus suillus 228
bats 19, 25, 43–4, 90–107
  big long-nosed 90
  bulldog 95–6, *96*
  bumblebee 93
  Californian leaf-nosed 96
  disc-winged 97–8, *98*
  eastern pipistrelle *103*
  European horseshoe 105
  European noctule 102
  false vampire 93–4, *94–5*,
      99

fisherman 95, *96*
free-tailed 99–100, *100*
fringe-lipped 107
funnel-eared 96
ghost 93
Gould's long-eared *102*
gray 90
hairless 100
hammer headed *92*, 107
hare-lipped 95
hispid 93
hog-nosed 90, 93
hollow-faced 93
Honduran tent bat *103*
horseshoe 94, *107*
little brown 105
long-eared *91, 98*, 107
mastiff 95
Mexican free-tailed 100,
      100, 105
mouse-eared 99, *105*
mouse-tailed 93, *93*
naked-backed 96, 99
New World bulldog 99
New World leaf-nosed 96,
      97, 99
New World sucker-footed
      97
New Zealand short-tailed
      91, 99
Old World leaf-nosed 95
Old World sucker-footed
      98
orange horseshoe *99*
painted 98
Seychelles sheath-tailed 90
sheath-tailed 93
slit-faced 93
smoky 97
spear-nosed long-tongued
      101
spotted 98, *99*
sword-nosed 97
tent-building 97
vampire 90, *91, 97*, 100–1,
      105
vespertilionid 98–9, *99*
white 96
wrinkle-faced 97
Wroughton's free-tailed 90
see also flying foxes; fruit
      bats
bears 20, 27, 138–9, *138–9*
  American black 139
  ant 19
  Asian black *138*, 139
  Asiatic black 134
  black 40
  brown 139
  grizzly 139
  Kamchatkan 139
  Kodiak 139
  polar 40, 134, 138, *138*,
      139
  sloth 134, 138, 139
  spectacled 134
  sun 139
  see also pandas
beavers 39, 40, 214, 215, *215*,
      216

behavior 33–7
Bettongia 58
bettongs 58, 63
bilbies *54, 55*, 61, 63
Bimana 130
bison 39, 40, 42, *42*, 192, 207
Bison bonasus 192
blackbucks 34, *34–5, 209*
Blarina brevicauda 81
boars *10–11, 196*, 197
bobcats 40
body structure 17, 18
bonbons 128
bongos 207
bonteboks *210*
borhyaenids 26
Bos sauveli 192
Bovidae 206
bovids 192, 197, 206, *206–11*
Bradypodidae 68
Bradypus torquatus 66
brains 17
Bubalus bubalis 192
bucks 206
buffalo 192, 195, *207*
bulls, blue 206
Bunolagus monticularis 229
Burramys parvus 57
burros 176
bushbabies 109, 110, 111, *133*

C

Caenolestidae 53
caenolestids 26
Callicebus 115
Callimico 115
Callithrix
  C. flaviceps 108
  C. pygmaea 115
Callorhinus 150
  C. ursinus 134
Calomys callosus 220
Caluromys 53
Camelidae 192
camels 32, 39, 192, 199, *199*
Camelus bactrianus 192
Canidae 20, 136
canids 136–7, *136–7*
Caniformia 135
Canis 136, 137
  C. adustus 137
  C. aureus 137
  C. latrans 134
  C. mesomelas 137
  C. rufus 134
  C. simensis 134, 137
Capra
  C. falconeri 192
  C. walia 192
Capromyidae 227
Capromys pilorides 227
capuchins 108, 115
capybaras 45, 214, 226, *226*
caracals *149*
caribou 31, *202*
Carnivora 20, 21, 25, 134
carnivores 26–7, 134–55
carnivorous marsupials 54–5
Castor 214
  C. canadensis 215, 216

C. fiber 216
Catagonus wagneri 192
Catarrhini 108
cats 20, 27, 34, 144–9, *149*
  see also cheetahs; jaguars;
      leopards; lions;
      tigers
cattle 206, *207*
  see also bison; buffalo
Cavia porcellus 225
cavies 225, 226
Caviidae 225
Caviomorpha 215, *215*, 225
caviomorphs 225
Cebus xanthosternos 108
cecum 62
Cephalogale 138
Cephalorhynchus heavisidii 156
Ceratotherium
  C. simum 174, 183
  C. simum cottoni 183
  C. simum simum 183
Cercartetus 56–7, 61
Cercocebus 117
Cercopithecidae 117
cercopithecids 117, 118
Cercopithecoidea 108
cercopithecoids 108
Cercopithecus 117
  C. cephus 117
  C. nictitans 117
  C. pogonias 117
  C. preussi 108
Cervidae 192, 201
Cervus alfredi 192
Cetacea 21, 156
cetaceans 158, 160
Chaetomys subspinosus 225
Chaetophractus nationi 66
chalicotheres 26
chamois 34
chausingas 207
cheetahs 30, 41, 134, *147*,
      147–8
cheirogaleids 109
Cheirogaleus 109, 112, *133*
Cheiromeles 100
chevrotains 192, 200, *200*
Chimarrogale hantu 74
chimpanzees 36, 108, 120,
      128–9, 128–31
Chinchilla brevicaudata 214
chinchillas 214, 226
Chinchillidae 226
chipmunks 217
Chironectes minimus 53
Chiroptera 19, 21, 90
Chlamyphorus
  C. retusa 72
  C. retusus 66
  C. truncatus 66, 72
Chlorotalpa tytonis 74
Chrotogale owstoni 145
Chrysochloridae 77
Chrysospalax trevelyani 77
Chysocyon brachyurus 137
civets 134, 144, *145, 145*
classification 19–21, 53
cloven-hoofed mammals 192
coatis 134, *140*, 140–1

Coelodonta antiquitatis 185
Coendou 225
Coleura seychellensis 90
Colobidae 117
colobids 117, 118–20
Colobus 118
    C. guereza 118–19
    C. satanas 119
colocolos 53–4
colugos 28, 88
common names 19
Condylura cristata 82
conservation 44–5
convergent evolution 32, *55*
coruros 227
cottontails 229
cougars *endpapers,* 40
cows 33
coyotes 134, 136, 137
coypus 226
Craseonycteridae 93
Craseonycteris thonglongyai 90, 93
Creodonta 134
Cretaceous period 22, 24, 27, 48
Cricetinae 222
Cricetomys 222
Crocidura 74
    C. russula 81
Crocuta 147
    C. crocuta 146
Crossomys moncktoni 214
Cryptoprocta ferox 134
Ctenodactylidae 227
Ctenomyidae 227
Cuon 136
    C. alpinus 136
cuscuses 28, 52, 56, *57*, 62, 63
Cyclopes didactylus 68
Cynictis penicillata 146
Cynocephalus
    C. variegatus 88
    C. volans 88
Cynodontia 23
cynodonts 23–4
Cynogale bennettii 134
Cynognathus 23
Cynomys 217
Cystophora cristata 153

D

Dactylopsila 56
Damaliscus hunteri 192
Daptomys 219
Dasypodidae 72
Dasyproctidae 226
Dasypus
    D. novemanctus 72
    D. pilosus 66
Dasyuridae 54
dasyurids 54, 61, 62, 63
Dasyurus viverrinus 64
Daubentonia madagascariensis 108, 110
deer 26, 27, 40, *200–3, 201–3*
    antlered 192
    brocket 192
    fallow *200*
    lesser mouse 192
    mouse 192, 200
    mule 194
    musk 192

New World 201
Old World 201
Pere David's 192
Philippine spotted 192
pudu *202*
red *202*
roe 41, *202*
wapitis *201*
water 201
    see also moose; muntjacs; reindeer
degus 227
Delanymys brooksi 222
Delphinidae 156
delphinids 156
Dendrogale
    D. melanura 86
    D. murina 86
Dendrohyrax 189
    D. validus 188
Dendrolagus 58
Dendromus 222
Dermoptera 21, 88
desert mammals 31–2
Desmana moschata 84
desmans 74, *75,* 84–5
Desmodontidae 97
Desmodus rotundus 91, 96
dholes 136, 137
Diaemus youngi 96
diana guenon *116*
Dicerorhinus sumatrensis 174, 185
Diceros
    D. bicornis 174, 183
    D. bicornis bicornis 184
    D. bicornis michaeli 184
    D. bicornis minor 184
Diclidurus 93
Didelphidae 53
didelphids 26
Didelphis virginiana 53
Dimetrodon 22, 23
dingos 27
Diphylla ecaudata 96
Dipodomys 218
    D. ingens 214
diprotodonts 56, 61
Distoechurus pennatus 57
dogs 20, 27, 33, 34, 36, 75, 135–7, *135–7*
    African hunting 136, *136,* 137
    African wild 134
    bush 137
    domestic 136
    raccoon 136, 137, *137*
    South American bush *136*
    see also foxes; prairie dogs; wolves
Dolichotis
    D. patagonum 226
    D. salinicolum 226
dolphins 156, *159,* 161, 163
    bottlenose 156, *161*
    Chinese river 163
    common 158, 161–2
    Ganges River 156
    Heaviside's 156
    Indus River 156
    Risso's 161
    river 156, *156*
    spinner 160

spotted 158
striped 158
    see also porpoises
dormice 224, *224*
douroucoulis *114, 132*
drills 108, 117
dromedaries *199*
Dromiciops australis 53
Dryomys nitedula 224
Dryopithecus 130, 131
Dugong dugon 164
Dugongidae 164
dugongs *164–5,* 164–6
duikers 206, *210*
dunnarts *63*
Dusicyon 137

E

echidnas 48, 49, 50–1, *50–1*
Echimyidae 225, *225,* 227
Echinosorex gymnurus 74, 79
echolocation 92, *106,* 106–7, 162
Ectophylla 96, 103
    E. alba 96
edentates 26, 27
Eira barbara 142
elands 45, 206
Elaphurus davidianus 192
elephant birds 40
elephant shrews 19, 74, 232–3, *232–3*
elephants *8–9,* 19, 25, 33, 39, 44–5, 167–73, *167–73*
Elephantulus 232
    E. refescens 233
Elephantus myurus 233
Elephas maximus 167, 173
Eliomys 224
elk 27, 40, 195, 203
Emballonuridae 93
embryonic dispause 65
endangered species 10, 38–45, 112
entellus 19
environmental degradation 40–1
Eocene period 90, 135, 164, 188, 192, 214, 225
eohippus 25
Epomops franqueti 107
Equidae 174
equids 174–9
Equus
    E. africanus 174
    E. burchelli 178
    E. caballus 174
    E. grevyi 174, 178
    E. hemionus 174
    E. przewalski 174, *176*
    E. zebra 174, 178
Eremitalpa granti 77
Erethizontidae 225
Erignathus barbatus 153
Erinaceus europaeus 77
ermines *1*
Eschrichtiidae 157
Eubalaena glacialis 156
Euderma maculatum 98
Eulemur 109, 112
Eumetopias jubatus 134, 150
Euoticus elegantulus 110
Eupleres goudotii 134

Eutheria 20
evolution 130–1
extinction, of species 38–9

F

falanoucs 134, 145
families 19
fanalokas 145
Felidae 20, 147
Feliforma 144
felines 147
Felis
    F. planiceps 147
    F. viverrina 147
ferrets 38, 142
flying foxes 43–4, *44,* 90, *92,* 105
forest mammals 28–9
fossas 134, *145*
fossils
    apes 130–1
    bats 90
    hyracoids 188
    insectivores 74
    mammals generally 20, 22
    mammoths 167
    Tenrecidae 76
foxes 40, 136–7, *137*
    Arctic 19, 31, 137
    bat-eared 136, 137, *137*
    fennec 136–7
    gray *137*
    North American gray 136
    red 19, 20, *20*
    see also flying foxes
fruit bats
    Barbados *101*
    Bulmer's 90
    cusp-toothed 90
    epauletted *92, 104,* 107
    Gambian epauletted *92*
    long-tailed 92
    Old World 90, 92–3, 101
    Wahlberg's epauletted *104*
fur *see* hair
fur seals 134, 150–2, *151,* 155
Furipteridae 97

G

Galagoninae 110
galagos 110
Galemys pyrenaicus 74, 84
Galidia elegans 146
Galidictis grandidieri 134
gaurs 207
Gazella
    G. bilkis 192
    G. leptoceros 192
gazelles 41, 192, 194, 206, *209*
geladas *116,* 117
gemsboks *210*
genera 19
genets 134, *144,* 145
Genetta 145
    G. cristata 134
    G. genetta 145
Geogale aurita 77
Gerbillinae 222
Gerbillus 214
gerbils 32, 214, 222, 223, *223,* 224
gerenuks *208–9*

gibbons 108, 120, 122–3, *122–3,* 129, 130, 131
Giraffa camelopardalis 192
giraffes 192, *193, 204,* 204–5
Giraffidae 192, 204
gliders 61, 63
    feathertail *56,* 57, 61, 62
    greater 56, *57,* 62
    mahogany 52
    squirrel 57
    sugar 34–5, 56, 62, *62,* 65
    wrist-webbed 56
    yellow-bellied *29*
Glis glis 224
gluttons *142*
Glyptodon 26, 27
gnus *194–5,* 206
goat-antelopes 206
goats 206, *211*
    see also tahrs
golden moles 74, 77, *77*
gophers 30, 218
Gorilla 130
    G. gorilla 108, 126
    G. gorilla beringei 127
    G. gorilla gorilla 127
    G. gorilla graueri 127
gorillas 44, *45,* 108, 120, *120,* 126–7, *126–7,* 130–1, *130–1*
Gracilinanus aceramarcae 52
Grampus griseus 161
Graphiurus 224
grassland mammals 30
Great American Interchange 27
guanacos 199
guenons 117
guinea pigs 225
Gulo gulo 134, 141
gundis 215, 227
Gymnobelideus leadbeateri 56
gymnures 74, 77, 79

H

habitats 28–32, 40–1, 73, 85
hair *16,* 16–17
Halichoerus grypus 153
hamsters 36
Hapalemur 109, 112
    H. aureus 112
    H. simus 108, 112
haplorrhines 108, 133
Haplorrhini 108, 111
hares 31, 218, *218,* 226, *226,* 229–31, *230–1*
hartebeests 206
hearing 17, *17*
hedgehogs 74, 77–9, *78–9*
Heliophobius argentocinereus 214
Helogale parvula 146
Hemicentetes semispinosus 77
Hemiechinus 78
Hemigalinae 145
herbivores 27
Herpestes
    H. auropunctatus 134
    H. sanguineus 146
    H. urva 146
Herpestidae 146
Hesperomyinae 218, 219
hesperomyines 218

Heterocephalus glaber 227, 228
Heterohyrax 189
　H. brucei 188
hibernation 62, 78, 102
Hippopotamidae 192
Hippopotamus amphibius 192
hippopotamuses 7, 40, 192, *197*, 198–9
Hipposideridae 95
hogs 27, 192, *194*, 197
Hominoidea 108, 130
hominoids 108, 120, 122
Homo 130
horses 26, 27, 174–6, *174–6*
humans 26, 108, 130–1
hutias 214, 227
Hyaena 147
　H. brunnea 146
　H. hyaena 146
Hyaenidae 20
Hydrochoeridae 226
Hydrochoerus hydrochaeris 214, 226
Hydromalis gigas 164
Hydromyinae 218
Hydromys chrysogaster 214
Hydropotes inermis 201
Hydruga leptonyx 150, 154
hyenas 20, *135*, 146–7
Hylobates 130
　H. agilis 123
　H. concolor 108, 123
　H. hoolock 123
　H. klossii 123
　H. lar 122, 123
　H. moloch 108
　H. syndactylus 123
Hylomys
　H. parvus 74
　H. suillus 79
Hypsignathus monstrosus 107
Hypsiprymnodon moschatus 58
Hyracoidea 21, 188
Hyracotherium 25
hyraxes 188–9, *188–9*
Hystricidae 227
Hystrix 227

I

ibex 192
Ichthyomys 214, 219
Idiurus
　I. macrotis 218
　I. zenkeri 218
impalas *12–13*, 37
indri 108, 110, *110*, 112, *120*
Indri indri 108, 110
Indricotherium 26, *26*, 183
infrasonic communication 172
Insectivora 21, 74
insectivores 74–85
Isoodon 55
ivory 172–3, *173*

J

jackals 20, *135*, 137
jaguars 40, 149
jaguarundis *149*
jaw development 15–16
jerboas 32, 214, 222–3
jirds 32
Jurassic period 22

K

kangaroos 27, 30, 58, *58–60*, 60, 62, 64, 65
　eastern gray 64
　giant 27
　Goodfellow's tree *59*
　musky rat-kangaroo 58, *59*
　rat-kangaroos 58, 60, 61, 63
　red 52, *59*, 64, *64*
　tree 58, 60, 62, 63
　western gray 58
Kerivoula picta 98
kinkajous 140, 141, *141*
klipspringers *210*
koalas 28, 29, 56, *56*, 62, 63, 65
koupreys 192
kowari 65
kudus 206
kultarrs 54
Kunsia 219

L

Lagidium 226
Lagomorpha 21, 229
Lagostomus maximus 226
langurs 108, 119–20
Lasiorhinus krefftii 52
Lavia frons 94
lemmings 31, *31*, 221
Lemmus lemmus 221
Lemur 109
　L. catta 109
lemurs 25, 108, 109–10, *109–11*, 112
　black 112
　broad-nosed 112
　broad-nosed gentle 108
　dwarf 109, 133
　flying 88, *88–9*
　gentle 109
　giant 27, 40
　gliding 28
　golden bamboo 108, 112
　gray mouse *111*
　hairy-eared dwarf 108, 112
　mouse 109, 112
　red mouse 108
　ring-tailed 109, *109*, 112
　ruffed 108, 109, *110*
　sportive 109–10
　white-rumped black 108
　see also aye-ayes
Leontopithecus
　L. caissara 108
　L. chrysopygus 108
　L. rosalia 108
leopards 134, 148, 149, *149*
Lepilemur 109, 112
Leporidae 229
Leptonychotes weddelli 154
Leptonycteris nivalis 90
Lepus 230
　L. americanus 231
　L. capensis 231
　L. europaeus 229
　L. saxatilis 231
　L. timidus 229
Liberiictus kuhni 134
limbs 17

Limnogale mergulus 74
Linh-guongs 192
linsangs *144*
lions 30, *33*, *36*, 134, *148*, 148–9
Lipotes vexillifer 156
litopterns 26
llamas 27
Lobodon carcinophagus 154
Lophiomyinae 222
Lophiomys imhausii 218, 222
Lophocebus 117
Lophuromys flavopunctatus 214
Loridae 109, 110
lorids 110
Lorinae 110
lorises 109, 110
Loxodonta africana 167, 173
Lutra felina 134
Lutreolina crassicaudata 53
Lycaon 136
　L. pictus 134
Lynx pardinus 134
lynxes 134

M

Macaca 117, 118
　M. fascicularis 118
　M. fuscata 108, 118
　M. maura 108
　M. mulatta 118
　M. pagensis 108
　M. silenus 108
　M. sylvanus 118
macaques *4–5*, 108, *116*, 118, *118*
macaws 40
Macroderma gigas 94
Macropodidae 58
Macropus 62
　M. giganteus 64
　M. robustus 63
　M. rufus 52, 64
Macroscelidea 19, 21, 232
Macroscelides 232
　M. proboscideus 232
Macroscelididae 74, 232
Macruromys elegans 214
maintenance behavior 33–4
Malacomys 214, 219
Mallomys rothschildi 215
Mammalia 14, 20, 21, 49
mammary glands 16
mammoths 27, 39, 167
Mammuthus primigenius 167
manatees 164–6, *166*
mandrills *116*, 117
Mandrillus 117
　M. leucophaeus 108
mangabeys 117
Manidae 212
Manis 212
maras 226, *226*
markhors 192
Marmosa 53
　M. andersoni 52
　M. robinsoni 65
marmosets 37, 61, 108, 115, 123, *132*
Marmosops handleyi 52
Marmota 217
marmots 216, 217, *217*
marsupial mice *see* dunnarts

marsupial moles 55, *55*, 61
Marsupialia 20, 21, 52
marsupials 18, 20, 24, 26, 27, 52–65, *52–65*
martens 141, 142
Martes 141
　M. pennanti 142
mastodons 39
Mayermys ellermani 214
meerkats 18, *46–7*, 146
Megachiroptera 90, 92–3, 101, 106–7
megachiropterans *92*, *104*
Megadermatidae 93
Megalonychidae 68
Megamuntiacus vuquangensis 192
Megatherium 26
Meles meles 143
Melinae 141
Melursus ursinus 134, 138
Mephitinae 141, 143
Mesocapromys
　M. auritus 214
　M. nanus 214
　M. sanfelipensis 214
Mesozoic period 24
Metatheria 20
Miacidae 135
mice 28, 33, 218–20, *219–20*, 222, 224, *224*
　Arabian spiny 214
　Australian hopping 32
　birch 224
　Delany's swamp-mouse 222
　European harvest *219*
　four-striped grass 220
　harvest 214–15
　house 219, *224*
　jumping 214, 224
　kangaroo 218
　one-toothed shrew 214
　palm 214
　pocket 215, 218
　pygmy 219
　spiny 219–20
　see also dormice
Microcebus 109
　M. myoxinus 108
microchipterans *94–5*
Microchiroptera 90, 92, 93–100, 101, 106
microchiropterans *91*, *101*
Microdipodops 218
Micromys minutus 214–15
Micropotamogale lamottei 74
Microtinae 220
microtines 221
Microtus pennsylvanicus 221
minks *16*, 134, 141, 142
Miocene period 26, 118, 153, 164, 218
Miopithecus 117
Mirounga
　M. angustirostris 154
　M. leonina 134, 154
moles 74, 77, *77*, 82–4, *82–5*
　see also marsupial moles
Molossidae 91, 99
Monachus
　M. monachus 134, 154
　M. schauinslandi 134, 154

M. tropicalis 154
mongooses 30, 34, 75, 134, 144, 146, *146*
monito del monte 53–4
monkeys 27, 108
　baboons 30, 34, 37, 108, 117, *117*
　black and white colobus *119*
　black colobus 119
　black snub-nosed 108
　capuchins 108, 115
　Central American squirrel 108
　colobids 117, 118–20
　crested guenon 117
　diana *116*
　drill 108, 117
　geladas *116*, 117
　golden snub-nosed *119*, 120
　greater spot-nose 117
　grizzled leaf 108
　guenons 117
　Hombold's woolly 115
　howler 40, 115
　indri 108, 110, *110*, 112, *120*
　langurs 108, 119–20
　long-tailed tree-monkey 68
　macaques *4–5*, 108, *116*, 118, *118*
　mandrills *116*, 117
　mangabeys 117
　mantled colobus 118–19
　mantled guereza *119*
　marmosets 37, 61, 108, 115, 123, *132*
　moustached 117
　New World 26, 28, 108, 111, *114*, 114–15
　night *114*, 115, *132*
　Old World 108, 117–20, 120
　pig-tailed snub-nosed 108
　Preuss's 108
　proboscis 108, *119*, 120
　red howler *114*, 115
　red uakari *114*
　rhesus 118
　silvery leaf-monkey 120
　snub-nosed 108, *119*, 120
　South American howler 34
　spider *28*, 115
　squirrel 115
　swamp 117
　tamarins 40, 45, 108, *114*, 115
　titis 115
　Tonkin snub-nosed 108
　woolly 115
　woolly spider *114*
　yellow-tailed woolly 108, 115
　see also apes
Monodelphis 53
Monodontidae 156
Monotrema 21
Monotremata 48
monotremes 17, 18, 27, 48–51, *48–51*, 64
moonrats 74, *75*, 77, 79

moose 194, 202
morganucodontids 24
Morganucodontid 24
Mormoopidae 96
Moschidae 192
mountain lion *endpapers*
mouse *see* mice
Mungos mungo 146
Mungotinae 146
munquis 114
Muntiacus annamensis 192
muntjacs 192, 200
Muridae 214, 218
Murinae 218, 222
murines 218, 219, 220
Mus 219
    M. musculus 219
Muscardinus avellanarius 224
musk deer *see* deer
muskrats 214, 220
Mustela
    M. felipei 134
    M. lutreola 134
    M. nigripes 142
    M. nivalis 134
    M. vison 141
Mustelidae 141
mustelids 141, 1–3, 153
Mustelinae 141
mustelines 142
musth 170
Myocastaridae 225
Myocaster coypus 226
Myomorpha 215, 215, 218
Myospalacinae 222
Myotis 99
    M. grisescens 90
    M. lucifugus 105
Myrmecobiidae 54, 55
Myrmecobius fasciatus 55
Myrmecophaga tridactyla 66
Myrmecophagidae 67
Mysateles garridoi 214
Mystacina tuberculata 91
Mystacinidae 99
Myzopodidae 98

N

Nasalis
    N. concolor 120
    N. larvatus 108, 120
Nasua nelsoni 134
Nasuella olivacea 141
Natalidae 96
Neofelis nebulosa 134, 149
Neolithic period 206
Neomys fodiens 81
Neophoca cinerea 150
Neophocaena phocaenoides 156
Nesolagus netscheri 229
Nesomyinae 222
Nesophontes 75
ningaui 52
Ningaui timaeleyi 52
Noctilio leporinus 95
Noctilionidae 95
noctules 102
nomenclature *see* common
    names; scientific names
noolbengers 28
Notopteris 92
Notoryctes typhlops 55
Notoryctidae 55

numbats 54, 54, 55, 61
nyalas 192
Nyctalus noctula 102
Nyctereutes procyonoides 136
Nycteridae 93
Nycteris grandis 93

O

ocelots 149
Ochotona
    O. daurica 229
    O. helanshanensis 229
    O. hyperborea 229
    O. koslowi 229
    O. pusilla 229
    O. rufescens 229
Ochotonidae 229
Octodontidae 227
Odobenidae 150
okapis 205, 205
Oligocene period 115, 138,
    152, 218, 225
olingos 140
Ommatophoca rossi 154
Ondatra zibethicus 214, 220
opossums 27, 61, 65, 108
    American 53
    lutrine 53
    mouse 53
    New World 28
    pale-bellied 65
    short-tailed 53
    shrew 53
    Virginian 53, 65
    water 52
    woolly 52
orang utans 33, 108, 120,
    124–5, 124–5, 129,
    130–1
orders 19
Oreonax flavicaudus 108
organs 18
Ornithorhynchus anatinus 48
Orthogeomys grandis 218
Orycteropodidae 190
Orycteropus afer 19, 190
Oryctolagus cuniculus 229
oryx 32, 32, 45, 192
Oryx dammah 192
Oryzomys 219
Otariidae 135, 150
otariids 150, 155
Otocyon megalotis 136
Otomops wroughtoni 90
Otomys 222
otters 40, 134, 141, 143
oxen 31, 206, 211
Oxymycterus 219

P

pacas 45
Pachydermata 168
Paguma larvata 145
Paleocene period 66, 190
Pan 130
    P. paniscus 108, 128
    P. troglodytes 108, 128
    P. troglodytes schweinfurthii
        128, 129
    P. troglodytes troglodytes
        128
    P. troglodytes verus 128
pandas 20, 134, 138–9, 139

pangolins 28, 212–13, 212–13
panther *endpapers*
Panthera
    P. leo 134
    P. onca 149
    P. pardus 148
    P. tigris 134, 148
    P. unica 149
pantherines 147, 148
Papio
    P. cynocephalus 117
    P. hamadryas 117
Paradoxurinae 145
Paraechinus 78
parental care 33
peccaries 192, 194, 196–7,
    199
Pectinator spekei 215
Pedetes capensis 218
Pentolagus furnessi 229
Perameles 55
    P. bougainville 52
Peramelidae 55
peramelids 55
Perissodactyla 21, 174
perissodactyls 26
Perognathus flavus 215
Peromyscus 219
Peroryctidae 55
Petauridae 56, 61
Petauroides volans 56
Petaurus 56, 61, 63
    P. breviceps 56
    P. gracilis 52
Petrodromus 232
Petrogale persephone 52
Phalanger 56
Phalangeridae 56
Phaner 112
Phascogale 54
phascogales 54, 54
Phascolarctidae 56
Phascolarctos cinereus 56
Phataginus 212
    P. gigantea 212
    P. temmincki 212
    P. tetradactyla 212
Phloeomys cumingi 219
Phoca
    P. caspica 153
    P. fasciata 153
    P. groenlandica 153
    P. hispida 153
    P. largha 153
    P. sibirica 153
    P. vitulina 153
Phocarctos hookeri 150
Phocidae 135, 150
phocids 150, 153–5, 154
Phocoena sinus 156
Phocoenidae 156
Pholidota 21, 212
Phyllostomidae 96, 101
Phyllotis 219
Physeter catodon 156
Physeteridae 156
pichiciegos 66
pigs 45, 192, 194, 196, 196–7
    see also boars; guinea pigs;
        hogs
pikas 229, 229
pinnipeds 150, 154
pipistrelles 103

placental mammals 18, 24, 26,
    27
Planigale 54
planigales 54
Platanista
    P. gangetica 156
    P. minor 156
Platanistidae 156
platypuses 20, 27, 48–9,
    48–50
platyrrhines 114, 115
Platyrrhini 108, 111
Pleistocene period 27, 39, 174,
    219
Pliocene period 66, 214, 218,
    225, 226
polar bears *see* bears
pollution 41
Pongidae 130
Pongo 130
    P. pygmaeus 108, 124
    P. pygmaeus abelii 124
    P. pygmaeus pygmaeus 124
porcupines 27, 28, 225, 225,
    227, 227
porpoises 156, 161
    see also dolphins
possums 61
    Australian 28
    brushtail 56, 62
    honey 28, 57, 61, 61, 62,
        64, 65
    Leadbeater's 56
    pygmy 56–7, 61, 62, 63,
        65
    ringtail 56, 62
    rock ringtail 56
    scaly-tailed 56
    striped 56, 57, 61
    see also opossums
Potoroidae 58
potoroos 52, 58
Potorous 58
    P. gilbertii 52
    P. longipes 52
Potos flavus 140
pottos 110, 111
prairie dogs 30, 217, 217
Praomys natalensis 220
prehistory 22–7
Presbytis 120
    P. comata 108
Primates 108
primates 14, 21, 26, 28, 33,
    34, 108–33
Priodontes maximus 66, 72
Proboscidea 19, 21, 167, 168
proboscideans 167
Procapra przewalskii 192
Procavia 188, 189
Procaviidae 188
Procolobus 118
Procyon
    P. lotor 140
    P. pygmaeus 134
Procyonidae 20, 140
procyonids 140
pronghorns 192, 195, 205, 205
Pronolagus crassicaudatus 229
Propithecus 110
    P. tattersalli 108, 112
Proteles cristatus 147
Prototheria 20, 21

Pseudocheiridae 56, 61
Pseudocheirus dahli 56
Pseudois schaeferi 192
Pseudonovibos spiralis 192
Pseudopotto martini 110
Pseudorca crassidens 161
Pseudoryx nghetinhensis 192
Pteralopex atrata 90
Pteronura brasiliensis 134, 143
Pteropodidae 92
Pteropus 105
    P. rodricensis 90
    P. vampyrus 90
Ptilocercinae 86
Ptilocercus lowii 86
pudus 202
puma *endpapers*
Pygathrix
    P. nemaeus 108, 120
    P. roxellana 120

Q

Quadrumana 130
quolls 54, 64, 65

R

rabbits 27, 229–31, 230–1
rabies 105
raccoons 20, 134, 140, 140
Ramapithecus 130–1
rats 28, 75, 218–22, 219–22
    African maned 218
    African mole-rats 222
    African swamp 214, 219
    American fish-eating 214
    Asian bamboo 222
    Australian water 214, 218
    black 220
    blind mole-rats 222, 228
    brown 219
    cane 227
    Cape dune mole-rat 228
    crested 222, 222
    Cuming's slender-tailed
        cloud 219
    earless water 214
    giant 222
    house 214
    kangaroo 32, 214, 218
    mole-rats 30, 222, 227,
        228, 228
    Mt Isarog striped 214
    multimammate 220
    naked mole-rat 227, 228,
        228
    Nile 220, 224
    paramo 219
    Peter's arboreal forest 214
    roof 219
    root 222
    ship 220
    silvery mole 214
    smooth-tailed giant 215
    speckled harsh-furred 214
    spiny 225, 225, 227
    swamp-rats 214, 219, 222
    western small-toothed 214
    zokors 222
    see also muskrats
Rattus
    R. exulans 220
    R. norvegicus 214, 219, 220
    R. rattus 219, 220

reindeer 31, 41, 193, 195, *202,*
203
reproduction 18, 36–7
Rhabdomys pumilio 220
Rheomys 219
Rhinoceros
R. sondaicus 174, 185
R. unicornis 174, 185
rhinoceroses 174, *182–6, 183–7*
African *182–3, 183–5*
Asian *184–5,* 185
black 174, 183, *183,*
184–5, *186,* 187
hook-lipped 183, *183*
Indian 174, *184,* 185, 186
Javan 174, 185, *185,* 186
square-lipped *182,* 183
Sumatran 174, *184,* 185,
186
white 174, 179, *182,*
183–4, 185, 186,
187
woolly 185
Rhinocerotidae 174
Rhinolophidae 94
rhinolophids 94
Rhinolophus ferrumequinum 105
Rhinopithecus
R. avunculus 108
R. bieti 108
Rhinopomatidae 93
Rhizomyinae 222
Rhynchocyon 232, 233
R. chrysopygus 232
ringtails *141*
roan 206
Rodentia 19, 21, 214
rodents 19, 25–7, 30, 32,
213–28, *215*
Romerolagus diazi 229
rorquals 157
Rousettus 92, 106

S

Saguinus oedipus 108
saigas *209*
Saimiri 115
S. oerstedii 108
sakis 115
Salpingotulus michaelis 214
Sarcophilus harrisii 54
Scandentia 19, 21, 74, 86
Scarabidae 190
scent glands 17
scientific names 19
Sciuromorpha 215, *215*
Sciurus carolinensis 217
Scotinomys 219
sea cows 164–6, *164–6*
sea otters *143*
sealions 134, 150, *150*
seals 33, 35, 134, 136, 150,
153–5, 153–6
see also fur seals; sealions
sebaceous glands 16–17
Selevinia betpakdalensis 224
Semnopithecus entellus 119
senses 17
sheep 192, 194, 206, *211*
shrews *35,* 74, 75, 76–7, 80–1,
80–1, 85
see also elephant shrews;
tree-shrews

siamangs *122,* 123
Sicista 224
sifakas 108, 110, 112, *113, 133*
simakobu 120
Simias 120
S. concolor 108
Simiiformes 111
Sirenia 21, 164
sirenians 164, 165, 166
Sivapithecus 131
skulls 15–16
skunks 141, 143, *143*
sloths 27, 28–9, *29,* 66,
68–70, *68–71,* 73
smell 17
soalas 192
social behavior 34–7
Solenodon 85
S. cubanus 75
S. paradoxus 75
solenodons 49, *74,* 74–6
Solenodontidae 74
Soricidae 80
sousliks 217
South American mammals
26–7
Spalacinae 222
Spalacopus cyanus 227
sparring 195
species 19
Speothos venaticus 137
Spermophilus 217
Spilocuscus rufoniger 52
springhaas 218
squirrels 215–17, *216–17*
African ground 146, 214,
217
African pygmy 216
African scaly-tailed 28
antelope ground 32
Columbian ground *217*
Eurasian red *216*
flying 28, 216–17
giant black *216*
grey 217
ground 30, *30,* 146, 214,
217, *217*
least *216*
scaly-tailed 218
southern *216*
see also chipmunks;
gophers; marmots;
sousliks
stoats *see* ermines
strandings, by whales 163, *163*
strepsirhines 108, *133*
Strepsirhini 108, 109
Suidae 192
Suncus 81
S. etruscus 74, 80
Suricata suricata 146
Sus
S. cebifrons 192
S. salvanius 192
Sylvilagus
S. dicei 229
S. idahoensis 229
S. insonus 229
synapsids 23

T

Tachyglossus 50
T. aculeatus 48

Tadarida brasiliensis 100, 105
tahrs *211*
Talpa
T. caeca 83
T. europaea 82
Talpidae 82, *82–3*
Tamandua
T. mexicana 67–8
T. tetradactyla 68
tamanduas *66,* 67–8
tamarins *40,* 45, 108, *114,* 115
Tapiridae 174
tapirs 27, 174, 180, *180–1*
Tapirus
T. bairdii 174, 180
T. indicus 174, 180
T. pinchaque 174, 180
T. terrestris 180
tarsiers 108, 111, *111,* 120
Tarsipedidae 57
Tarsipes rostratus 57
Tarsius 111, 133
T. bancanus 111
Tasmanian devils 53, 54
Tasmanian tigers *54,* 55
taste 17
Taterillus gracilis 224
Taxidea 141
Taxidea taxus 143
Tayasuidae 192
tayras 142, *142*
teeth 15–16
Tenrec ecaudatus 77
Tenrecidae 76
tenrecs 74, *75,* 76, 76–7, 85
Tertiary period 24–5, 26, 27,
193, 200, 204
Thamnomys rutilans 214
Therapsida 23
therapsids 15, 23
Theria 20
Theropithecus gelada 117
Thomasomys 219
Thryonomys 227
T. swinderianus 227
thylacines 27, 54, *54,* 55, 61
Thylacinidae 54
Thylacinus cynocephalus 55
Thylacosmilus 26
Thyropteridae 97
tigers 27, 39, *39,* 134, 148,
148
titis 115
toddy cats 144
Tolypeutes tricinctus 66, 72
touch 17
Trachops cirrhosus 107
Trachypithecus 120
T. auratus 120
T. cristatus 120
T. delacouri 108
T. francoisi 120
T. geei 120
T. pileatus 120
Tragelaphus buxtoni 192
Tragulidae 192
Tragulus javanicus 192

tree shrews 19, 86–7, *86–7*
Tremarctos ornatus 134
Triassic period 10, 24
tribosphenic mammals 27
Trichechidae 164
Trichechus 164
T. inunguis 164
T. manatus 164
T. senegalensis 164
Trichosurus 56
Trichys lipura 227
Tubulidentata 19, 21, 190
tuco-tucos 30, 227
Tundra mammals 30–1
Tupaia 86
T. chrysogaster 86
T. glis 86, 87
T. longipes 86
T. montana 87
T. nicobarica 86
T. palawanensis 86
Tupaiidae 86
Tupaiinae 86, 87
Tylomys 219

U

uakaris 115
uintatheres 25
Uintatherium 25
Uncia uncia 134
ungulates 30, 35
even-toed 192–211
odd-toed 174–87
Urocyon 136
U. cinereoargenteus 136
Uroderma 96, 103
Urogale everetti 86
Ursidae 20
ursids 138, 139
Ursus
U. americanus 139
U. arctos 139
U. maritimus 134
U. thibetanus 134, 139

V

Vandeleuria oleracea 214
vaquitas 156
Varecia 109
V. variegata 108
Vespertilionidae 98
vespertilionids 98–9
vibrissae 16, 17
vicuñas *198,* 199
viscachas 226, *226*
vision 17, 132–3, *132–3*
Viverra 145
V. civettina 134
Viverravidae 144
Viverricula 145
viverrids 144
Viverrinae 145
voles 220–1, *221*
Vombatidae 56
Vulpes 19, 136, 137
V. vulpes 19, 20, *20*
V. zerda 137

W

wallabies 30, 58, 62, 65
large-eyed forest *59*
rock-wallabies 52, 58,
58, 63
wallaroos 63
walruses 150, *152,* 152–3
wapitis *201*
weasels 34, 134, 136, 141,
142
whalebone 158
whales 25, 33
baleen 157, 158, 160, *160,*
161, 162
beaked 156, 161
blue 156, *159*
bowhead 156
false killer 161
fin 156
gray 157
humpback *2–3,* 42–3,
43, 158, 162, *162,*
163
killer 156, *157,* 158, 160,
161
minke *163*
northern right 156
right 157
rorquals 157
sei 156
sperm 156, 160, 161, *163*
toothed 156, 158, *160,*
161, 162
whalebone 157
white 156
whaling 43, 163, *163*
wildebeests 45, *194–5, 210*
wolverines *40,* 134, 141, 142,
142
wolves 20, *40,* 136, 137
Ethiopian 134
maned 30, *30, 41,* 41–2,
136
red 134, *136*
wombats 56, 62, 63, 64, 65
giant 27
northern hairy-nosed 52
Wyulda squamicaudata 56

X

Xenarthra 21, 66
Xerus 146, 217
X. erythropus 214

Y

yaks *206*
yapoks *52,* 53, 61

Z

Zaglossus
Z. bruijnii 48
Zalophus californianus 150
zebras 174, 178–9, *178–9*
Zenkerella insignis 218
Ziphiidae 156
zokors 222

ACKNOWLEDGMENTS

The editors and publishers would like to thank the following people for their assistance and support: Dr Sue Hand, University of New South Wales, Australia; Dr Merlin Tuttle, Bat Conservation International Inc., USA; Maureen Colman; Helen Cooney; Robert Coupe; Lesley Dow; Vanessa Finney; Jane Fraser; Tristan Phillips; and Meryl Potter.

COPLEY LIBRARY

DEMCO